ANCIENT FISHING AND FISH PROCESSING

BLACK SEA STUDIES

2

The Danish National Research Foundation's
Centre for Black Sea Studies

ANCIENT FISHING AND FISH PROCESSING IN THE BLACK SEA REGION

Edited by
Tønnes Bekker-Nielsen

AARHUS UNIVERSITY PRESS

ANCIENT FISHING AND FISH PROCESSING IN THE BLACK SEA REGION
Proceedings of an interdisciplinary workshop on marine resources and trade in fish products in the Black Sea region in antiquity, University of Southern Denmark, Esbjerg, April 4-5, 2003.

Cover design by Jakob Munk Højte and Lotte Bruun Rasmussen

Mosaic with scene of fishermen at sea from a tomb in the catacomb of Hermes in Hadrumetum (Sousse Museum, inv.no. 10.455). Late second century AD. 320 x 280 cm. Photo: Gilles Mermet.

Printed in Gylling by Narayana Press
ISBN 87 7934 096 2

AARHUS UNIVERSITY PRESS
Langelandsgade 177
DK-8200 Aarhus N

73 Lime Walk
Headington, Oxford OX2 7AD

Box 511
Oakville, CT 06779

www.unipress.au.dk

The publication of this volume has been made possible by a generous grant from the Danish National Research Foundation

Danish National Research Foundation's
Centre for Black Sea Studies
Building 328
University of Aarhus
DK-8000 Aarhus C
www.pontos.dk

Contents

Illustrations and Tables

Athena Trakadas: The Archaeological Evidence for Fish Processing in the Western Mediterranean

Vladimir F. Stolba: Fish and Money

Jakob Munk Højte: The Archaeological Evidence for Fish Processing in the Black Sea Region

Bo Ejstrud: Size Matters

Introduction

> ... like some blameless king, who upholds righteousness,
> as the monarch over a great and valiant nation: the earth yields its
> wheat and barley, the trees are loaded with fruit, the ewes bring
> forth lambs, and the sea abounds with fish by reason of his virtues,
> (Homer, *Odyssey* 19.110-114, trans. Samuel Butler)

To any reader familiar with Classical literature, lambs, fruit and ears of corn are familiar symbols of prosperity and fertility. But fish? It would seem that to the hero – and the author – of the *Odyssey*, an abundance of fish was a characteristic of the good city-state and a testimony to the virtue of its ruler.

The Danish city of Esbjerg boasts a fishing port as well as an equestrian statue of the virtuous King Christian IX in the main square. These two features alone, then, would qualify it as the venue for a conference on ancient fishing. In addition, the city is home to a branch campus of the University of Southern Denmark, one of the partner institutions in the Danish National Research Foundation's Centre for Black Sea Studies. It was at the Esbjerg campus that the papers in this volume were first presented in the course of a workshop in April, 2003. Some twenty researchers took part in two days of lively discussions ranging as widely as the travels of Odysseus.

Historians, classicists and archaeologists dealt with the question of ancient fish processing from the viewpoint of their disciplines, but in addition, we were fortunate to have an inspiring presentation on "The biochemistry of fish processing" by Hans Otto Sørensen, biochemist and laboratory manager at Triple Nine Fish Protein, Esbjerg. As the world's second largest producer of fish protein concentrate, Triple Nine undertakes extensive research into the biochemistry of fish processing and fish spoilage. We regret that it was not possible to include Hans Otto Sørensen's presentation in this volume.

After the conference, it was felt that it would be useful to complement the papers with a survey of the prehistory of fishing in the northern Black Sea region. Nadežda Gavriljuk kindly undertook to write a chapter on this subject at short notice.

For the ancient world, much of our information on fish in general is derived from the extensive range of sources dealing with fish as a foodstuff and, from the time of Hippokrates (c. 400 BC) onward, with the medicinal properties of fish. John Wilkins' survey of the textual evidence reveals that among ancient authors – that is to say, among those whose works have been preserved entirely or in fragments – some possessed a detailed knowledge of numerous types of fish, their origins and their taxonomy. When assessing the wealth of detailed information and culinary advice presented by, e.g., Athenaios, one should keep in mind, however, that the opportunity to pick and choose from a wide variety of fish species was open only to affluent and sophisticated members of the elite, such as Athenaios himself. In this respect, the work of Galen may provide a more realistic impression of ancient dietary habits.

The literary sources for processed fish products such as *garum* are supplemented by a large and growing volume of epigraphical and archaeological evidence, but it is remarkable that – as pointed out by Robert Curtis in his chapter on "Sources for Production and Trade of Greek and Roman Processed Fish" – not until the twentieth century were serious attempts made to relate the testimony of the texts to the archaeological material, or to view them in the light of parallels with contemporary fish processing in Southeast Asia (where fish sauces similar to the ancient *garum* are produced today). In fact, it was not until the 1960's that the first large-scale systematic study relating the different source categories to each other (Ponsich & Tarradell 1965) was published.

Literary texts and inscriptions of the Roman period indicate that in their time, fish processing took place along the coasts of the empire from the Atlantic to the Black Sea, and archaeological finds corroborate the testimony of the textual sources. They also indicate that the consumption of *garum* and related products was more widespread, in spatial and social terms, than that of fresh or salted fish. As evidenced by the presence of *garum* amphorae in landlocked Augusta Raurica, discussed by Bo Ejstrud in this volume, fish sauces could be transported far inland and were – unlike fish – generally not expensive. A possible modern parallel is tomatoes: in winter, these are still something of a luxury in northern Europe, but tomato ketchup is not.

One aspect of fish as food that was not dealt with in any of the presentations but taken up in the discussions was the definition of "fresh" fish in antiquity. As pointed out by Hans Otto Sørensen, attitudes to freshness in foodstuffs are largely culturally determined. Fish at a certain stage of incipient spoilage might be rejected in one culture, but considered edible (or even preferable to the fresh article) in another. The popularity of *garum* suggests that compared with modern southern Europeans, ancient Romans had a greater tolerance towards strong smells in fish (and no doubt elsewhere in their daily environment) and thus might be less inclined to reject fish on olfactory criteria alone; on the other hand, the rarity of references to food poisoning in the literature and the practice of night fishing suggest that the ancients' criteria for freshness were not too far from our own – but the topic deserves further research.

In her survey of the archaeological evidence for fish processing in the western Mediterranean, Athena Trakadas focuses on the remains of salting installations in present-day Morocco, Portugal and Spain. Studying large processing complexes such as Lixus or Cotta can offer valuable clues for interpreting similar installations in the Black Sea area, e.g. at Tyritake, Myrmekion or Chersonesos. It is notoriously difficult to estimate production capacity from the remains of salting vats, or to relate potential capacity to actual production, but a rough comparison of the cubic capacity at western Mediterranean salteries with those of the Black Sea salteries may nonetheless provide a useful basis for comparison.

Trakadas also discusses the question of origins: did fish processing take place in the Punic period, or was it introduced to the western Mediterranean by the Greeks? Scholars of an earlier period, when diffusionism and Greek cultural superiority were taken for granted, favoured the notion that fish processing was a Greek innovation, but the archaeological evidence points to Punic origins.

The numerous and well-documented fish processing sites in the west permit certain generalizations about chronology and spatial location. Athena Trakadas identifies common factors determining the location of processing sites: proximity to the sea, plentiful freshwater resources and salt; also fuel for concentrating liquids through evaporation. Kilns for producing amphorae are often found in conjunction with processing sites, and these in turn again require fuel as well as suitable clay. The most important input is obviously fish, and in the last analysis, large-scale fish processing depends on the ability of the fishermen to supply large quantities of suitable species. It has been claimed (Gallant 1985) that ancient sea fishing technology was inherently inefficient, but starting from the most important literary source, the *Halieutika* of Oppian (second century AD), Tønnes Bekker-Nielsen demonstrates that the ancient fishing gear – which would include lines with multiple hooks and seines worked by two teams of fishermen – was sufficiently advanced to produce sizeable catches of fish for salting or *garum* production. In fact, the most important constraint on the further expansion of the ancient fishing sector was the inability to conserve fish for any length of time, a constraint that could be partly overcome by drying, smoking or salting fish, or by converting them into fish sauce.

The theme of fishing productivity is continued by Anne Lif Lund Jacobsen's paper on the use of modern fishery statistics as an indication of ancient catch sizes in the same waters. Jacobsen has worked with fishing statistics in the early modern period and warns against simplistic assumptions that present catches will correspond to actual or potential catches in history. The potential productivity of a fishery is determined by a number of interrelated factors: the state of the marine ecosystem, human exploitation of fish stocks, efficiency of the fishing gear. Because environmental factors play such a large part, fishing statistics from, e.g., Malaysia, are irrelevant to fishing in the Mediterranean or Black Sea. She identifies a number of other problems inherent in such diachronic comparisons and also points out that (contrary to the assumption underlying T.W. Gallant's analysis of nineteenth and twentieth-century catch statistics, which led him to conclude that the average catch would hardly feed the fisherman and his family) the fish consumed on board or in the fisherman's household are not normally included in the official catch figures. Fishing statistics thus indicate the marketable surplus after the needs of fishermen and their dependents have been met, not the total harvest from the sea.

With Nadežda Gavriljuk's contribution, our geographical focus shifts to the northern Black Sea region and the indigenous nomads of the steppe zone. Generally "fish" and "sea" are not concepts associated with "steppe" or "nomad"; the extent and importance of pre-Greek fishing activity in the northern coastal regions of the Black Sea has been overlooked – and, Gavriljuk argues, underestimated. Fish motifs in Scythian ornaments indicate that fish had a place in the self-perception of the nomadic population, and the rivers of the steppe zone would offer excellent prospects for catching freshwater and migratory fish. Such fishing activities, however, took place within the *oikos* economy. Fishing on a larger scale is not, at present, attested before the late fifth and early fourth century, when we find evidence of fishing and fish processing on a substantial scale at Elizavetovka, a site that is discussed in greater detail by Jakob Munk Højte. On the lower Dnieper, fishing on a "commercial" scale is attested in the second century BC. Gavriljuk concludes that while "subsistence" fishing formed part of the nomad economy at an early date, cultural and commercial contacts with the Greeks were the driving force behind the transition to "commercial" fishing.

The iconographical theme is taken up again in Vladimir Stolba's survey of "Fish and money", demonstrating that fish species depicted on coins of the Pontic Greek cities are often realistically rendered and can be securely identified, the most popular being the various species of sturgeon. While this clearly testifies to a general interest in fish, Stolba warns against jumping to conclusions about the place of fish in the local economies – or to the absence of fishing in cities that do *not* strike coins with fish emblems. Fish and marine species have a vast range of potential symbolic, mythological and religious connotations, as evidenced by the "eagle and dolphin" emblem used, *inter alia*, by the cities of Sinope, Histria and Olbia (and as the emblem of the Danish National Research Foundation's Centre for Black Sea Studies).

From the abstract spheres of mythology and symbolism, we pass to the practical aspects of fish processing, as Jakob Munk Højte takes us on a guided tour of the salting facilities along the northern Black Sea coast. Many of the excavated fish processing tanks have been covered up to protect them from the rigours of the Pontic climate; and what appears to be the largest site, Chersonesos, has not yet been completely excavated. The dimensions of the processing complexes testify to the extent of the Crimean fish salting operations: the combined cubic capacity of the salting vats at Tyritake was 457 cubic metres, and for Chersonesos an estimated 2,000 cubic metres, whereas the largest of the western Mediterranean complexes, Lixus, had a combined cubic capacity of just over a thousand cubic metres. Unfortunately, as Højte points out, we do not know whether vats were always filled to capacity, nor how many times a year. It may be speculated that in the cooler climate of the Crimea, the annual number of *garum* production cycles would be less than in Spain or North Africa.

In chronological terms, too, comparison between fish-processing sites of east and west are instructive: Athena Trakadas found that fish processing in the western Mediterranean lagged in the second century BC but was revitalised under the early Empire while the Crimean salteries seem to have attained their maximum capacity in the second century AD.

Amphorae, the most common form of transport containers in the ancient world, form an important category of sources for economic history. Surprisingly, they offer very little positive evidence for Pontic fish processing before the Christian era. This paradox is examined in depth by John Lund and Vincent Gabrielsen: while we have textual evidence for the import of *tarichos* and other fish products from the Black Sea to the Aegean, no amphora types have been convincingly identified as containers for fish products. Taking the textual evidence for a Pontic salt-fish trade at face value, Lund and Gabrielsen offer two possible explanations for the absence of transport amphorae: either fish products were transported in re-used wine or oil amphorae (as may be the case with an amphora from the Varna shipwreck); or they were packed in other types of containers, for instance, baskets (for *tarichos*) or barrels (for *garum*). The notion of recycled amphorae is attractive but there are no ancient parallels for large-scale re-use of amphorae in this manner (in contrast to the re-use of individual amphorae for a plethora of domestic purposes). Therefore with our present state of knowledge, the latter hypothesis appears the most likely.

"Vanished" amphorae also form the subject of Bo Ejstrud's chapter on "Estimating trade of wine, oil and fish-sauce", but in his case wine, not fish is missing from the equation. Taking the large and well-documented body of amphora fragments from Augusta Raurica (modern Augst in Switzerland, near Basle) as a starting point, Ejstrud first discusses the relationship between amphora finds and actual volumes, then turns to the relative importance of wine, oil and fish-sauce. Even when allowance has been made for the different size of the containers, the volume of fish-sauce attested at Augst is remarkably large in relation to the amount of wine. Since there is no good reason why consumption patterns in the Roman veteran colony at Augst should differ from comparable settlements elsewhere in the western Empire, the conclusion that a considerable amount of wine remains undocumented imposes itself. Presumably the inhabitants of Augst did not only import wine in amphorae but also in other containers, such as barrels.

The lesson to be learned from the contributions by Lund, Gabrielsen and Ejstrud is that while the importance of amphorae as archaeological source material remains undisputed, focusing on this one category of containers may in fact provide a distorted picture of commodity flows at a given site or date. As pointed out by Stefanie Martin-Kilcher (1990), it is quite possible that the dramatic drop in the number of *garum* amphora fragments in the course of the third century that can be observed at many central European sites, does not reflect a corresponding decline in consumption but a changeover from southern European suppliers to fish sauce produced in northern Gaul, which was shipped in barrels.

A similar problem is that of the missing salting tanks, discussed by Jakob Munk Højte. The Crimean fish-salting industry probably did not reach its maximum capacity until the late second century (at the same time as, or slightly later than, its Iberian and African counterparts) and no processing facility has been dated earlier than the first century AD. Yet fishing (attested by finds of fishing tackle) and fish processing (attested by literary sources) is known to have taken place – but where? One explanation could be that earlier processing facilities were primitive, along the lines of those found at Elizavetovka, and have been obliterated by later, more permanent structures. Fish salting originally took place within the *oikos*, and the diversification into market-oriented *garum* production requiring large capital outlays may have provided the impetus to relocate and concentrate activities in urban centres.

It also needs to be remembered that salting is not the only means of preserving fish; reducing the relative water content of the fish by smoking or drying will have similar effects to salting. At the Elizavetovka site, the remains of a smokehouse have been tentatively identified; as for drying, this requires little equipment except some wooden racks that would, at the most, leave a few post-holes. Fish drying requires no input of salt and produces a commodity that (unlike salt fish, which must be soaked and cooked) is immediately edible and can be used as animal fodder. In the twentieth century, dried fish was produced in large quantities in the North Atlantic (e.g., Newfoundland, Iceland). Great quantities were also produced in south Russia: it is reported that during a fuel shortage in the aftermath of the 1917 revolution, the Bolshevik authorities in Turkestan seized large stocks of dried sturgeon awaiting export and used them as locomotive fuel.[1] Since it leaves so few traces in the archaeological record, the extent of ancient fish drying is difficult to establish. Given the limited number of references to dried fish in the literary sources, it may primarily have been produced within the household, perhaps as a supplementary food in times of famine or poor fishing, or as a staple item of the lower-class diet – Lif Jakobsen reminded us that in some modern fishing communities, dried fish was considered "trash food". We might, however, also see fish drying as a first stage in a sequence of technological development and increasing market orientation: dried fish for the household economy – salt fish for the regional market – *garum* for the distant markets of the Roman Empire.

In the history of fish processing along the Black Sea, a key question concerns the genesis of the Crimean *garum* industry: whether its origin was supply-driven or demand-driven. Did Crimean fish salters turn to *garum* production in response to a glut of smaller fish that could not be sold as table fish or processed into *tarichos*; or did they diversify from their core operation to exploit a growing demand for a new culinary ingredient *à la mode*? The implications of this debate extend far beyond the kitchen table, since the two positions reflect two fundamentally different conceptions of the Roman economy. On the one hand, we have the "primitivist" view that Roman primary producers walked a thin line between prosperity and famine, adapting to the exigencies of a changing economic environment. In recent years, this has been combined with the environmental history approach focusing on natural phenomena such as epidemics and climate change to explain past economic behaviour. Within this frame of interpretation, the introduction of *garum* production could be local producers' response to a Pontic "fish bonanza" such as those that have occurred throughout history (the Baltic herring bonanza of the Middle Ages being one example). Recurrent windfalls of fresh fish might stimulate fish-salters to develop new products.

A "modernist" interpretation, on the other hand, would not necessarily look for the causal factor in the marine environment of the Black Sea, but view the introduction of *garum* as an active response by informed Crimean fish-salters to a growing market for *garum* in the Mediterranean world. The salting cisterns of Myrmekion or Tyritake were clearly not built by destitute fishermen, but by members of the elite with access to technology, capital and information about distant markets. By combining a cheap, plentiful – but perishable – commodity (fish) with an easily obtainable raw material (salt), resulting in a product with higher unit value and longer storage life (*garum*), they were able to exploit those markets.

The current stage of our knowledge favours the notion that the diversification from fish salting into *garum* production was driven by demand, but our picture of ancient fishing and fish processing along the shores of the Black Sea is far from complete. Future research may come up with different answers, even pose entirely different questions. The present volume is not intended as the last word on Pontic fish processing, and if it provokes academic controversy and scholarly discussion on its subject, it will have served its purpose well. It is hoped, however, that the surveys of source material and overviews of the *états des questions* provided in the individual contributions as well as the consolidated bibliography will serve as useful aids to future researchers in a field that still has much to offer.

University of Southern Denmark, Esbjerg
November 2003 *Tønnes Bekker-Nielsen*

Notes

1 Blacker 1922, 266; Brun 1930, 109.

Fish as a Source of Food in Antiquity

John Wilkins

1. *Introduction*

This chapter reviews literary evidence for the consumption of fish in antiquity and has two aims. The first is to complement the archaeological evidence presented in other chapters in this volume; the second is to comment on the nature of the literary sources, to show that they provide not merely "evidence" on the topic of salted fish but in addition a valuable cultural commentary on this type of food. This commentary reveals the variety of types available and the enormous range of interest in what might at first sight seem a simple foodstuff.

Sources on the ancient diet are diverse and potentially misleading. They tend to be influenced by strong literary and philosophical traditions which value morality over health and nutrition, and the care of the soul over care of the body. Two extreme illustrations may be found in Ovid's *Fasti* and Plato's *Republic*. At *Fasti* 6.169-86, the goddess Carna is said to be a goddess of traditional values whose festive foods are emmer wheat and pork. The foods the goddess is said to reject are imported fish and foreign birds such as the guinea fowl (which is said to have come to Rome from North Africa) and francolin (which is said to be imported from the Aegean). Rome, this text asserts, was not traditionally a fish-eating society, or at least did not construct herself as such.

Similarly, when describing the ideal diet for the rulers in the *Republic* (372), Plato's Sokrates sets out vegetarian foods that are based on agriculture as those which are most desirable. This privileging of cereals and meat over fish and imported foods is widespread and pervades many literary sources, including much of Greek comedy, which is one of the major literary genres to focus on the consumption of fish (see Wilkins 2000, ch. 6).

This ancient perspective has been reinforced by two modern works, Gallant 1985 and Davidson 1997. The first, which is discussed elsewhere in this volume, uses a statistical approach to claim that fish stocks made only a small contribution to the ancient diet, while the latter focuses on fish as a luxury product. Taken together, these claims suggest that the majority of people in the ancient Mediterranean area ate little or no fish. I argued in Wilkins 2000 and 2001 that Davidson's picture is partial (he is interested only in the wealthy) and that Gallant's case is misconceived since he sets the calorific value of fish in the diet against that of staple cereals rather than considering fish beside meat, vegetables and other complements to a cereal diet. I argued that fish was accessible to all members of the ancient polis, depending on cost and supply, both of which varied greatly. I also argued, incidentally, in Wilkins 2000, that Davidson was mistaken in assuming that the institution of the symposium was exclusive to the elites of the Greek cities. Davidson's argument on the role of desire in ancient Greek literature is an excellent case which maps on to the ancient diet (with particular reference to fish and wine) in a fascinating way; but it does not accurately reflect the ancient diet as currently understood.

I mentioned Ovid and Plato above because they are deeply embedded in Roman and Athenian culture. A number of the texts I am about to use do not derive specifically from either Athens or Rome and so reflect broader concerns in the ancient world. In the first place, my focus is on medicine, on the author of the Hippocratic text *Regimen II*, Diocles of Carystus, Mnesitheus of Athens, Diphilus of Siphnos, Hicesius of Smyrna and Galen of Pergamon. Secondly, I consider the sympotic author Athenaios of Naukratis. These authors cover a period from the fourth century BC (the author of *Regimen II*, Diocles and Mnesitheus) to the second/third centuries AD (Galen and Athenaios). These are familiar authors to the historian of ancient fishing. They are used extensively for example in Curtis 1991. What I undertake is to explore the importance of fish in these medical and sympotic authors in general, and not to use them merely as sources for vital pieces of evidence in a reconstruction of the ancient fish industry. My main headings will be: (a) geographical concerns; (b) fish in relation to other foods; (c) concerns of terminology and taxonomy; and (d) attempts to give an overview of the diet, from the imperial period in which Galen and Athenaios were writing, back over a millennium of Greek culture.

2. *Texts on fish*

2.1 *Medical texts: a brief survey*

The Hippocratic *Regimen II* (which probably dates to about 400 BC) has a substantial section on fish (48) dividing them according to dryness (*xerotatoi*) and lightness (*kouphoi*, largely the fish that inhabit rocky waters) and heaviness (*barus*, largely fish that inhabit rivers and marshes). Salted fish (*tarichos*) is "drying and attenuating" (Jones, *xerainousi kai ischnainousin*), fat ones are gently laxative, while the driest come from the sea, the moistest from lakes. The driest *tarichos* comes from the driest fish. This classification derives from the scientific categories of *Regimen 1*, in which it is argued that the cosmos, the world and human and animal life are based on the principles of fire and water.

Later in the fourth century, Diokles of Carystus seems to have discussed fish in at least two treatises, *Hygieina* and *Hygieina to Pleistarchus*. He too is interested in dryness (Diokles, fragments 225-27, van der Eijk) and in soft or tough flesh in fish. For *tarichos*, he is interested in fat and non-fat fish that have gone into the pickle. There is little to distinguish Diocles from his predecessor in respect of fish, other than to say that more fish were probably included in the discussion. There is thus a possibility that the Hippocratic list was expanded.

This feature is more marked in Mnesitheus (fourth century BC) and Diphilus (third century BC), while Hicesius of Smyrna addresses the *chule* or juice that the food brings with it or causes to be produced in the body, and the qualities of a food that might affect the stomach. These may be new features, but it is hard to say given the fragmentary nature of the evidence. All three expand what the Hippocratic author had said.

Galen, writing in the second century AD has more to add. He has a major geographical concern, on which more shortly; he expands on the environmental concerns of the Hippocratic author; and he also rejects the scientific basis of *Regimen 1*, even denying that treatise Hippocratic authorship.

Of the medical authorities mentioned above, most of the fragments of works on fish by Mnesitheus, Diphilus and Hicesius survive only in quotations in the *Deipnosophistae* of Athenaios, a slightly younger contemporary of Galen.[1]

2.2 *Athenaios: a brief summary of the Deipnosophistai*

The *Deipnosophistae* is a sympotic text in the tradition of Plato's *Symposium*. Unlike that work and nearly all later symposia, food is at the centre of the *Deipnosophistai* and the familiar idea that wine provokes thought and philosophy – unlike food which impedes thought and discussion – is turned on its head. The fifteen books follow the order of the meal, and the symposiasts debate with each other over the correct way to approach their material. They recline in Rome, the centre of the world, and bring the fruits of research in libraries to the dinner table of Larensis, a minor Roman magistrate. They are clients of the great patron, and have an ambiguous relationship with Rome. But the text does not seem to adopt the hostile approach to fish which is frequently found in Roman authors, for example in the ninth book of Pliny's *Natural History* (9.53). Rather, at the beginning of the fish section we are told how

> Thereupon, slaves entered bearing an enormous quantity of fish from sea and lake, on silver platters, so that we marvelled at the luxury as well as at the wealth displayed. For our host had brought everything but the Nereids. And one of the parasites and flatterers remarked that Poseidon must have sent the fish to Nittunius; not however through the agency of the merchants in

> Rome who sell a tiny fish for a huge price; rather he must have brought them himself, some from Antium, others from Taracina and the Pontian islands opposite, still others from Pyrgi which is a city of Etruria. For the fishmongers of Rome do not fall short, even by a little distance, of those who were once satirised in Athens (Athenaios, 6.224b, trans. Gulick).

A little later, at the beginning of the next book, Athenaios moves from sympotic conversation to an alphabetical list of fish, which raises questions of (a) composition and (b) the ordering of his material. This makes for difficult reading, but if we set form aside the data provided is invaluable.

I want to turn to two representative passages: one is medical from Galen, the other is "sympotic" from Athenaios. I shall then draw out what I think are four important features from them.

2.2.1 *On grey mullet*

> The grey mullet belongs to the family of scaly fish that grows not only in the sea but also in pools and rivers. This is why the various grey mullet differ greatly from one another, so that the class of sea mullet appears to be another one from that in the pools, rivers or swamps, or in the drains that clean out the city latrines. ... They are better or worse according to their food. For while some have plenty of weed and valuable roots and so are superior, others eat muddy weed and unwholesome roots. And some of them that dwell in rivers running through a large town, eating human dung and certain other such bad foods, are worst of all ... It is also clear from what has been said that, for this reason too, one sea is better than another so far as it is either completely clear or receives many large rivers like the Pontus. For in such a sea the fish are as superior to those living in pools as they are inferior to those in the open sea. ... Now this mullet like any other sea fish does not possess many small spines. But the mullet that enters the sea from rivers and marshes is full of such spines, much the same as other fish from the same source. ... Some of our own people call the fish produced in rivers "white mullet", believing that they are a different species from grey mullet. ... This fish is also one of those that is pickled, and the variety from pools becomes much improved when prepared in this way. For it gets rid of everything in the taste that is slimy and foul-smelling. The recently salted fish is superior to the one that has been pickled for a longer time. But a little later there will be a general discussion about pickled fish, as also about fish that can be kept

in snow until the next day (Galen, *On the Properties of Foodstuffs* 3.24 = 6.708-13 Kühn, trans. Powell)

Now Athenaios:

> As we ate our salt fish many of us had a desire to drink. And Daphnus, raising his hands, said: Heracleides of Tarentum, my friends, says in his work entitled *Symposium* that a "moderate quantity of food should be eaten before drinking, and chiefly the dishes that form the ordinary courses at the beginning of the feast. For when foods are served after an interval of drinking, they counteract what settles on the stomach from the effects of wine and becomes the cause of gnawing pangs. Some even think them unwholesome – I mean the different kinds of green vegetables and salt-fish – possessing as they do a pungent quality. ... Diphilus of Siphnos says that salt-fish, whether from sea or lake or river, has little nourishment or juice; it is dry, easily digested, and provocative of appetite. The best of the lean varieties are cubes, *horaia*, and the like; of the fat, the tunny steaks and young tunny. When aged, they are superior, being more pungent, particularly the Byzantian sorts. The tunny steak, he says, is taken from the medium-sized young tunny, the smaller size resembling the cube tunny, from which class comes also the *horaion*. The Sardinian tunny is as large as the tuna. The mackerel is not heavy, but readily leaves the stomach. Spanish mackerel is rather purgative and pungent and has poorer flavour, but is filling. Better are the Amynclanian and the Spanish sort called Saxitanian, which are lighter and sweeter. Now Strabo, in the third book of his *Geography*, says that Sexitania, from which this fish gets its name, is near the Isles of Heracles, opposite New Carthage, and that there is another town called Scombroaria [Mackerel town] from the mackerel caught there. ... The river crow-fish from the Nile, which some call crescent, but which among the Alexandrians is known by the special name of "half-salt" is rather fatty, quite well-flavoured, meaty, filling, easily digested and assimilated, and in every way superior to the mullet. ... (Athenaios 3.120b-121c, trans. Gulick).

The main features that emerge in Galen's account of the grey mullet, an important fish in the salting industry, are: the difficulty in distinguishing one species from another; the crucial role played by habitat; and his assimilation of much detail from varied sources. Athenaios, by contrast, quotes from sources and does not synthesise them into a smooth account. He refers to more places (but Galen refers to Spain elsewhere (3.30), on salted tuna), covering

Spain and Egypt as well as the Black Sea. He also draws heavily on medical authors – Heracleides and Diphilus – and reflects the integration of comment on medicine and comment on eating and symposia. He too is interested in varieties and the differences either between fish or between different ages and cuts of fish. We can pick out generic features from these passages, which are brief extracts from much more extensive comments on fish and salt fish. Galen has fifteen chapters on fish, a number of which include salted fish, and one on salted food in general (3.23-37 and 41); Athenaios discusses salt fish explicitly at 3.116a-121e, and in passing in many other passages.

3. *Concerns of Galen and Athenaios*

3.1 *Geographical concerns*

Galen's review of food in general in *de alimentorum facultatibus* (On the Properties of Foodstuffs) is comprehensive. He lists cereals in more detail than does Athenaios, plants in similar detail, meat and fish in less detail. Both authors range over many geographical areas, from Spain to Syria and the Black Sea to North Africa, but they concentrate on Asia Minor and Alexandria; in Athenaios' case there is much on Athens and the cities of mainland Greece; in Galen the Greek mainland is completely ignored, in favour of the greater Roman Empire. Here are some details on the Pontic region, which is the main focus of this volume. In his survey of all foods, Galen refers to specific places 73 times. Of this number, he refers to places related to the Pontic region 18 times (to Pontus 3 times, to Bithynia 3, to Dorulaion, Juliopolis, Claudiopolis, Crateia, Nicaea, Prusa, Cappadocia and Pamphylia once, to Thrace twice, to Phrygia twice. Additional references to Asia are normally to Mysia (3) and Galen's home town of Pergamon (2). This focus on place is invaluable. Galen has modified the Hippocratic focus on the environment (most notably in *Airs, Waters and Places*), though those concerns are still evident in the discussion above on the grey mullet in relation to rivers, deltas and different seas. To this environmental interest, Galen adds specific places that he has visited, or so at least he implies. Autopsy is one of his main modes of research. He thus gives excellent regional variation of diet, particularly for cereals but also for fish.

Athenaios can match Galen and outstrip him. Place is a vital consideration for Athenaios, and he provides the most specific data on the ancient diet that is available. He refers to so many hundreds of cities that I do not give figures. His data is normally also indexed by time, from the date of the author quoted. I'll return to Athenaios on geography.

3.2 *Fish in relation to other foods*

There are large differences. Galen has three books, one devoted to cereals, one to other plants and one to animal and fish products. Fish is a smaller component, comprising 15 out of 147 chapters. It is clear in Galen's mind that cereals far outweigh fish in importance. Athenaios reverses this emphasis. He, by contrast, has 15 books, the three central of which (6-8) are largely devoted to fish, while, in addition, shellfish and salted fish are treated in book three. Just a small part of the evidence of salted fish in book three is quoted above. What are the implications of this coverage? One answer is that Athenaios, unlike Galen, who in his travels often notes what peasants and other poor people eat, is concerned with the life of luxury and all the variety and distinction that money can buy. The perspective of luxury explored by Davidson 1997 is certainly relevant here.

3.3 Terminology and Taxonomy

But so is another perspective. Fish provide as great a challenge to the taxonomer as any division of ancient food. Athenaios attempts to cover a large number of them, while Galen does the same for plants, and to some extent for fish, as we saw above. I discuss this further in Powell (2003) ix-xxi. Here are some representative passages:

Athenaios on the *hepatos* or *lebias*; Athenaios goes to a doctor and two philosophers to try to pin down the names of this fish (or fishes):

> Diocles says that this is one of the rock fishes. Speusippus says that the *hepatos* is like the *phagros*. According to Aristotle it is solitary, carnivorous and has jagged teeth. (Athenaios, 7.301c).

> Galen on *seris*:

> there is another kind of wild herb which is less cutting than those mentioned; this kind appears to belong between the two, having neither a definitely cutting nor a thickening effect. The general name for these is *seris*; but the individual species are given different names by rustics, such as lettuce, chicory, the Syrian *gingidia* and countless similar ones in every region. The Athenians use the term *seris* indiscriminately for all of them; for the ancients did not allot any names to the individual species. (Galen, *Thinning Diet* 3).

Galen here identifies a developing problem for the taxonomer, which was not noticed by "the ancients" but now requires attention.

Galen on firm-fleshed fish finds errors in his sources, as does Athenaios on occasion:

> Phylotimos also wrote about these in the second book of *On Food* as follows: weevers, pipers, sharks, scorpion fish, horse mackerel

> and red mullet [a long list follows]. This is Phylotimos' statement. But let us examine each individual item mentioned, from the beginning. Now weevers and pipers, to those who have eaten them, clearly have firm flesh. But there is no one species of shark. For the fish that is highly prized among the Romans, which they call *galaxias* belongs to the family of sharks [*galeoi*]. This fish does not seem to occur in Greek waters, which is why Phylotimos also appears to be unaware of it. ... It is clear that the *galaxias*, which is in high repute among the Romans, is one of the tender-fleshed; but the other sharks are firm-fleshed (Galen, *On the Properties of Foodstuffs* 3.30).

Galen is making various claims, to be more thorough, more accurate, more up to date and more accurate in his taxonomies.

However, the clear message for us is that there was considerable uncertainty in the ancient world considering families of fish. While we can accept modern identifications of fish that are based on archaeological evidence, ancient evidence is more ambiguous partly because it does not rest on our "scientific" classifications and partly because so many varieties of fish were eaten, both fresh and salted.

3.4 Overview of the diet

Galen and Athenaios attempt to cover the whole Roman Empire, from Spain to Syria and beyond. They also cover a millennium of Greek culture, from Homer, whom both mention, to their own day.

Problems of taxonomy are lexical as well as botanical and zoological. This is why these two sources produce the detail they do on salted fish and grey mullet. They are indeed valuable texts.

These texts can be made to work for us in various respects and will produce various results.

Let us take first the example of *garum*. Athenaios does not mention *garum* very often. He has an entry on *garos* in his list of seasonings at 2.67b-c:

> FISH SAUCE (GAROS). Cratinus has this: "Your pannier will be chock full of fish-sauce". Pherecrates: "He has fouled his beard with fish sauce". Sophocles in *Triptolemus*: "the sauce made of pickled fish". Plato: "They will souse me and suffocate me in rotten fish sauce." That the noun is masculine is proved by the masculine article which Aeschylus uses when he says: "the sauce made of fish" (Athenaios 2.67b-c, trans. Gulick (adapted)).

These attestations are fascinating. They draw on the authors of Greek comedy (Cratinus, Pherecrates and Plato) and tragedy (or more likely satyr play), namely Sophocles and Aischylos. There was evidently clear knowledge of *garum* in fifth century Athenian drama. Was it widely used, as in the Roman period? It would appear not, for the term rarely appears in Athenaios and he had every reason to mention it and none to omit it. Furthermore, the early culinary text Archestratos' *Life of Luxury* does not mention it, but uses related flavourings based on salt water and vinegar, with reference also to *tarichos*. If *garum* was known but not widespread in the fifth and fourth centuries, why did Athenaios not tie it in to his own time, as he does sturgeon and many other items? Galen, in contrast, frequently refers to garum as a flavouring in his own time, as Curtis 1991 has observed.

The next promising area that might be explored is the role of salted fish in the "cutting" or remedying of thick humours. Phlegm is particularly in need of treatment in this respect in Galen's humoural medicine. It is clear from Galen's treatise *On the Thinning Diet* and from chapter 3.41 of *On the Properties of Foodstuffs* that salt and vinegar are probably as important in providing this effect as the fish itself. Athenaios' characters have a similar interest in vinegar, oil and salt at 9.384f, in a passage on acidity of taste and language which ties in with the passage before us in book 3 (cf. p. 25).

A third example is offered by the *Life of Luxury* of Archestratos. This is a mock epic poem of the fourth century BC that puts all the fish back into epic poetry that Homer had famously omitted from the *Iliad* and *Odyssey*. The very title *the Life of Luxury* is problematic for an author such as Athenaios, since it would appear to encourage immorality, as he often points out, but Archestratus also provides much that Athenaios values. This includes alternative names for fish; different species in different locations around the Mediterranean; different forms of preparation for the table. Archestratos thus addresses, some six centuries earlier, the geographical and taxonomic interests of Athenaios and Galen. Indeed, Athenaios sometimes ties comments made by Archestratos to the Roman period, as Galen did above on the *galaxias*. The following fragment of the poem gives a good example of Archestratos' use of detail, with reference to geography, type of fish and mode of salting:

> and a slice of Sicilian tuna < >
> cut when it was about to be pickled in jars (*bikoi*).
> But I say to hell with *saperde*, a Pontic dish,
> And those who praise it. For few people
> Know which food is wretched and which is excellent.
> But get a mackerel on the third day, before it goes into salt water
> Within a transport jar as a piece of recently cured, half-salted fish.

> And if you come to the holy city of famous Byzantion,
> I urge you again to eat a steak of peak-season tuna; for it is very good and soft.
> (Archestratos, fr. 39 Olson and Sens)

Conclusions

Athenaios shows us that there was an extraordinary range of interest in salt-fish and other fish in ancient texts. When thinking about the relationship between fish and food in ancient thought, Athenaios demonstrates that fish is a topic not just for zoological research (which Aristotle, for example, pursued in *On the History of Animals*), but for the symposium and even for the theatre.

Medical and culinary study of salt-fish and other fish in antiquity were not separated in ancient thought. Doctors were happy to write about the symposium and sympotic writers often refer to medical works.

Galen and Athenaios have very different objectives, but they share interests in the identification and taxonomy of fish.

For all his value, Athenaios is not interested in the majority of the population. It is in fact quite difficult to find detail on the poorer sections of society – but it is there, as I showed in Wilkins 2000 and Wilkins 2001. Galen was interested in poorer citizens, but only in respect of the plants and cereals that they consumed. Fish for him, as for many other texts, were the expensive single fish ("singletons") that were affordable only for the rich.

Galen and Athenaios offer a double perspective. They review their own period, with an extensive overview of the second and third centuries AD. But they also provide on an encyclopaedic scale reference back to specific datable authors. Each can bring benefits to Black Sea studies.

We should remember, finally, the fine fish-plates of Athenian and South Italian manufacture, many of which were sent to Olbia and subsequently excavated there. Many of them were painted by the same or similar vase-painters as the familiar drinking cups and mixing bowls of the symposium. Some thousand of these plates are known, with artistic representations of fish upon them that link them with the deipnon-symposium in a way similar to the link between wine and pottery and song. This is a further expression of the integration of fish into Greek culture.

Notes

1 See Athenaios *Deipnosophistai* 3.116e and 118b-c for Hicesius, 120e for Diphilus, 121d for Mnesitheus. Also 8.355a-358c. For Mnesitheus, see also Bertier 1972, 30, 178-9, 190-1, 194-205.

Sources for Production and Trade of Greek and Roman Processed Fish

Robert I. Curtis

The workshop sponsored by the Danish National Research Foundation's Centre for Black Sea Studies coincides with a heightened scholarly interest in ancient foods of all kinds and thus presents an excellent opportunity to review the present state and future directions of the study of the production and trade of Greco-Roman processed fish, salt-fish (*salsamentum* or τάριχος) and the fish sauces (*garum*, *liquamen*, *allec*, and *muria*).[1] Focus on Black Sea products seems quite appropriate for two reasons. First, development of the fishery resources of the Black Sea may have been a prime motivation behind Greek colonization of the region, perhaps as early as, or before, the seventh century BC.[2] And second, the earliest modern study devoted to the ancient processed fish industry (Köhler 1832) focused on its manifestation along the coastal areas of the Black Sea. I wish to look in some detail at the sources available for studying the production and trade in processed fish products, not only in the Black Sea area but also throughout the Greco-Roman world. Before doing so, however, I would like to comment briefly on why I think this workshop takes place at a key juncture in the study of this important aspect of ancient life during the classical period. Over nearly the last half-century study of the production and trade in fish by-products has shown a marked vitality. It was not always so.

1. *History of research*

Production and commerce in processed fish continued after the fall of the Roman Empire, though the degree of their practice varied greatly from one geographical area to another and from one period to another. Though salt-fish products, particularly *garum*, continued to receive mention during the medieval and early modern periods – in both the East and the West, in letters, literary works, government documents, and the like – the knowledge of and interest in ancient fish products became primarily the province of physicians and scientists.[3] This was no doubt a result, for varying reasons, of the declining economic and social importance of these products in most places, and because of the prominence given to the medical works of Galen and Oribasios and to the encyclopedic *Natural History* of Pliny the Elder among doctors, naturalists, and ichthyologists. So, one thinks, for instance, of references to ancient fish products, particularly the fish sauces, in the sixteenth-century works of Francis Rabelais, Guillaume Rondelet, and Pierre Belon.[4] A thorough study of the place of processed fish products in the literary, social, and economic life of this period, however, remains a prominent desideratum.

The first serious, and still valuable, work devoted specifically to ancient fish products was entitled Τάριχος, *ou recherches sur l'histoire et les antiquités des pêcheries de la Russie méridionale*. Published in 1832 in St. Petersburg, Russia, under name of Köhler,[5] this monograph collected, for the first time, an impressive quantity of ancient literary sources devoted not only to identification of fish products and their uses but also to their manufacture. It is notable as well for its anthropological approach to the subject, since Köhler related the ancient products to those coming from contemporary Black Sea fish-salting factories. The groundbreaking work of Köhler, however, was not followed up, save in two short publications. In 1871 J.K. Smidth published, in Danish, a short article with the long title "Historical Observations on the Condition of the Fisheries among the Ancient Greeks and Romans, and on their Mode of Salting and Pickling Fish."[6] After briefly discussing ancient fishing and the fish available to the Greeks and Romans, he devotes a mere four pages to fish salting. He does, however, give some interesting tidbits on fish processing in the North Atlantic in the mid-nineteenth century. In 1892, Georg Eberl brought out a short monograph entitled *Die Fischkonserven der Alten*, devoted specifically to preserved fish, in which he concisely identifies various kinds of salt-fish and fish sauces.[7] Although more readily available, these works advanced our knowledge little beyond what Köhler had provided. The only other important 19th-century work significantly treating the topic was the 1890-dissertation, in Latin, by Paul Rhode who focused specifically on tuna fishing and its salted by-products.[8] In 1910, Robert Zahn published his Pauly-Wissowa article on *garum*, which, while collecting the ancient literary evidence, including many sources not utilised by Köhler, and citing several post-Roman sources, makes considerable use of painted inscriptions on amphorae.[9]

The works of Smidth, Eberl, Rhode, and, to a lesser extent, Köhler and Zahn, have one thing in common. Their sources are almost wholly literary in nature. By the middle of the twentieth century, however, important advances did come, particularly in our understanding of the technical characteristics of the fish sauces. For example, Pierre Grimal and Thomas Monod, in an important article published in 1952, related the nature of the ancient sauces, and how they were made, with that of the modern sauces of Southeast Asia, particularly those produced in Vietnam and Thailand. This was followed in 1961 by Claude Jardin's article, which, besides briefly noting the importance of underwater archaeology in studying ancient trade in fish by-products, includes a discussion of the chemical composition and nutritive value of the fish sauces.[10] In investigating the nutrition of fish sauce Jardin was among the first to employ in a serious way data from scientific disciplines, namely biochemistry and microbiology, to assist in understanding ancient fish by-products. Even so, with these two exceptions, discussions of ancient processed fish products throughout the first half of the twentieth century continued to rest primarily on literary and, to a lesser degree, epigraphic evidence. This situation substantially changed in 1965.

The first comprehensive study of the Roman salt-fish industry that made extensive use of archaeological evidence for the salteries themselves was that of Michel Ponsich and Miguel Tarradell, entitled *Garum et industries antiques de salaison dans la Méditerranée occidentale*.[11] It was not that no archaeological excavations of fish-salting factories had taken place before this time; rather, it was that few classical historians had paid much attention to publications of them. Archaeology remained essentially divorced from history and literature. Ponsich and Tarradell focused on the Roman industry as it operated in southern Spain and Portugal and across the Strait of Gibraltar in Morocco. Their work highlighted the importance of the physical remains of fish processing and defined the criteria used by subsequent scholars to identify similar installations elsewhere, such as in France and Tunisia.[12] In addition, Ponsich and Tarradell's work shed a bright light on *salsamentum* as a commercial product. Prior to that time the focus had been almost exclusively on the fish sauces, the product most often receiving comment in ancient literary sources. Their work revolutionised the study of the ancient fish salting industry and imparted a liveliness to the investigations that continue to this day.

Up to this time, most historians and literary critics consistently underestimated the social and economic value of fish and their by-products. They saw the fish sauces as expensive products whose major use was as a condiment for foods and whose medicinal value was minimal at best.[13] The pejorative characterization of the sauces in ancient literature (e.g. Seneca, *Letters* 95.25; Pliny, *HN* 31.93), resulting from their seemingly bizarre production process and their ill repute for strong smell and taste, influenced modern scholars not to take them seriously. "Our stomachs would probably revolt at a dish prepared with garum", was the conclusion of one mid-twentieth-century Italian scholar.[14] But, the upsurge in interest in them in the second half of the 20th century derives from a confluence of various scholarly approaches and technological advances, of which the work of Ponsich and Tarradell represents the beginning.

First of all, the past thirty-eight years have seen a significant increase in excavations, particularly in the Western Mediterranean, and a growing awareness of the historical value of archaeological evidence. Second, scholars have begun to see more clearly the importance of combining literary and historical evidence with the archaeological and epigraphic. In addition, the growing partnership between archaeologists and scientists has become particularly important as scientific instruments and analyses begin to play a more significant role in investigating the ancient material world. Third, while not ignoring the more spectacular archaeological projects and the traditional social and political subjects, scholars have turned more attention to ancient daily life, especially food. What ancient Greeks and Romans ate, how they cooked it, how it was processed, if necessary, what was involved in its transportation, where it was shipped, and who participated in all these activities have become questions of great import. Fourth, there has developed an increased focus by many scholars on the lower classes – freemen, freedmen, and slaves – of Greco-Roman society, the very ones who made up the population of those engaged in processing and trading fish and fish products.[15] Concomitant with this is an increased interest in non-elite populations in the provincial areas, particularly in regard to the influence the "other" had on Rome. And, finally, scholarly interest in the ancient economy, especially the role of the city in the economy, has increased significantly.[16] All of these elements have extensively augmented the source material available for a study of fish by-products far beyond what was accessible to Köhler, Smidth, Eberl, and Zahn, and have increased the number and types of questions demanding answers. I would like to survey those sources briefly, indicating what they can provide for our knowledge of ancient processed fish products. At the same time I hope to identify areas needing further investigation.

2. *Production*

Information about production of ancient processed fish products derives in the main from three types of sources: literary references, archaeological remains of salting installations, and comparative data from production methods of similar modern products. The fact that Greco-Roman literary references to processed fish products derive from many different sources and genres and, for the most part, that they are casual in nature, strongly imply that salt-fish and fish sauce were commonplace in ancient life. Of particular importance are the gastronomical works, such as the *Hedupatheia* of Archestratos (*fl.* fourth century BC), the *De re coquinaria* of Apicius (*fl.* first century AD), and the *Deipnosophistai* of Athenaios (*fl.* c. AD 200), which relay information on what fish products were eaten and how they were prepared.[17] References to fish products also come from drama, both comedy and tragedy, such as the extant works of Aristophanes and Plautus, and from Athenaios who preserves extracts from the works of many Greek dramatists, such as Nikostratos, whose works no longer exist. They come from the epigrams of Martial, the satires of Horace, and ancient scholia. They come from the didactic poetry of Manilios, from ancient letters, both literary and private, such as those of Seneca and Ausonius, and from the many papyri of Egypt, a source ignored by Zahn in his otherwise extensive 1910 Pauly-Wissowa article.[18] They come from the novel of Petronius and the oratorical work of Quintilian. They come especially from medical and veterinary treatises, such as those of Galen, Oribasios, Xenokrates, and Pelagonius, from the agricultural manuals of Cato, Varro, Columella, and Cassianus Bassus, and from the encyclopaedias of Pliny the Elder and Isidore of Seville. They even come from grammarians, from ancient glossaries, from the *Regulae*, or "Rules", of Christian monastic orders, and from the Jewish *Talmud*.[19]

The types of information derived from literary sources vary from the mundane, such as the proper spelling or gender of the terms for fish sauce, to the more important, such as the names of salt-fish that divulge something about their preparation and detailed directions for producing fish by-products.[20] So, for example, the only description for making *salsamentum* comes from the first-century AD agricultural treatise of Columella, and then only by indirection. After describing how to salt pork, by laying down alternating layers of meat and salt, he implies that the process is the same for salting fish (Columella 12.55.4). Other authors provide the names of fish by-products that hint at the type of fish used, the part of the fish chosen, the shapes into which the fish were cut, or their saltiness. For example, θυννίδες refers to tunny, ὑπογάστρια specifies the stomach portion, τετράγωνον indicates a rectangular-shaped piece, and ἡμιτάριχος states that the salt-fish is only half-salted. Interestingly, all specific terms for salt-fish are Greek; Latin expressions, where not subsumed under the general term *salsamenta*, are mere transliterations of the Greek, such as *trigonum*.[21]

Recipes for making fish sauce, however, are more numerous and come from different periods. The earliest descriptions are in the *Historia Naturalis* (31.93-95) of Pliny the Elder and in the *Astronomicon* (5.656-681) of Manilios, both of the first century AD. Two recipes of the third century AD are found in works attributed to someone else. These include the preparations (*confectiones*) of Ps.-Rufius Festus and of Ps.-Gargilius Martialis.[22] These ancient works provide information on ingredients used and their proportions, note various additives, sometimes specify the containers used, and hint at the process of manufacture. The fact that directions for producing fish sauce found in three post-classical sources closely coincide with earlier classical descriptions indicates that production methods changed little over the centuries. These post-classical sources include the seventh-century AD encyclopedia of Isidore of Seville (*Orig*. 20.3.19-20), a recipe appearing at the head of an eighth or ninth-century AD Merovingian manuscript of a medical treatise (Paris Bibl. Ms. Latin 11219), and the tenth-century AD Greek agricultural manual called the *Geoponika* (20.46.1-6). The latter source, which may ultimately derive from the sixth-century AD Latin work of Cassianus Bassus, provides the most detailed description extant for preparing *garum*.[23]

These recipe descriptions, in combination with other genres, such as satire, allow us to draw certain conclusions about the general character of these fish by-products. Manilios, for example, makes it clear that *salsamenta* and fish sauce can be by-products of the same production process, particularly where large fish, such as tunny and mackerel, are concerned. No part of the fish was wasted. Small, whole fish or the innards of large fish were ideal for producing fish sauce. Second, basic ingredients necessary to produce fish by-products include a supply of fish, salt, and fresh water. Third, salt-fish came in various shapes, and in appearance were probably rather coarse looking, particularly if salted with scales still attached, and shriveled up with a dry appearance, if heavily salted. Plautus (*Poen.* 240-244) tells us that, before eaten, salt-fish sometimes had to be washed with fresh water. Unfortunately, no ancient author describes in detail what fish sauce looked like, but descriptions of the production processes imply that *garum*, *liquamen*, and *muria* were salty liquids, and fairly clear, if strained, while *allec* was probably a rather thick salty mush containing scales, bones, and undissolved fish matter.[24] Odors from their production were no doubt strong, but probably not as bad as some authors, such as Martial (3.77.5) or Artemidoros (*Onirokritikon* 1.62), would have us believe when they describe *allec* and *garum* as "putrid".[25] More than this we cannot gather, but we can make reasonable conjectures from what we do know about modern fish sauces, a subject discussed below (p. 39).

Ancient authors are also excellent sources for identifying places for production. For example, the tenets of ancient dietetics stressed eating the correct foods to maintain the proper balance of humours. Foods had defined powers to promote good humours and to counteract imbalances. These powers, expressed in terms of opposites, such as heating and cooling, moist and dry, laxative and binding, and so on, were linked with time of year, geographic location, and other environmental variables. So, many medical writers when talking of the medicinal value of fish frequently stressed certain fish or fish by-products from specific locations. Among the most important of these sources, for example, is Xenokrates' *De alimentis ex fluviatilibus*, or "Food from Aquatic Animals", which dates to the first century AD.[26] In Books IV and V he discusses the dietetic value of salted fish from sea, river, and lake, and makes special note of the Spanish mackerel and the small tunny, or pelamys, of the Black Sea region. The second-century BC historian Polybios (4.38.4; 31.25.5) notes that salted fish was one of the major products the Pontic areas supplied both to the cities of Greece and to Rome itself. Galen (*On the Properties of Foodstuffs* 3.30.5), writing in the second century AD, while also remarking on these same fish products, praises the salt-fish of Sardinia as well. Gastronomic writers, in discussing particular foods, highlight delicacies from around the Mediterranean, while satirists and other critics focus on famous and expensive fish products. Archestratos (frg. 39 Olson and Sens), for example, praises the salted tunny of Sicily, while Martial (13.102) is one of many who praise *garum sociorum* from New Carthage in Spain. Geographers in describing different geographical locations often speak of the food resources of a particular region or city. Strabon, for example, describes the tunny watches of Italy (5.2.6, 8; 6.1.1) and North Africa (17.3.18), and the salting factories of Spain (3.1.8; 3.4.2, 6) and the Black Sea (7.6.2; 11.2.4; 12.3.1, 19). These differentiations are not casual. Today, various fish products from diverse regions of the world do have distinct tastes, colors, consistencies, and so forth, because not only fish but also processes can vary.[27]

The second category of sources for production of fish by-products are the numerous remains of salting installations discovered by archaeological excavations conducted in the Western Mediterranean and Black Sea regions. The best-documented area remains that part of modern Spain and Portugal that in the Roman period went under the name Baetica, the region that formed the object of Ponsich and Tarradell's 1965 book. In 1988 Ponsich updated that work by significantly increasing the number of Spanish and Portuguese sites discussed from 15 to 89.[28] Many of these had received mention in literary sources, while many others were recognised from the characteristic physical evidence. Signs that a salting installation operated at a particular location include, most prominently, salting vats (*cetariae*), usually square or rectangular in shape and varying in size and depth.[29] A waterproof coating (*opus signinum*) covered the interior walls and floor; the angles at the bottom were reinforced, and the floor also had a shallow cuvette to assist in cleaning. It has for a long time been assumed that production of fish sauce took place in small vats, usually round in shape and less deep than the larger rectangular ones assumed to have been devoted to making salt-fish (*salsamenta*). Excavations at Neapolis (mod. Nabeul), in Tunisia, however, have uncovered at least one large rectangular basin that contained bones of small fish, mainly anchovy and sardines, preserved whole. The identification of the product as the sauce *allec* seems secure. Evidently, large rectangular vats, not just small round ones, could also be utilised for fish sauce production.[30] This fact raises questions bearing on seasonality of production and on specialization of product. Excavators in Pompeii have found dolia containing the dried remains of *allec*, also made with anchovies.[31] Finds of amphorae, with shapes usually associated with salt-fish products, evidence for a salt supply (salt mines or flats), and a source of fresh water also characterise salting installations.[32]

Major salting installations discovered so far on the Mediterranean coast of Roman Baetica include Sexi (mod. Almuñecar), Malaca (mod. Malaga), and Carteia (mod. El Rocadillo). On the Atlantic coast, installations have been discovered at Baelo (mod. Belo), Barbate, Puerto Real, Gades (mod. Cádiz), Las Redes, and Cerro del Trigo. Salting installations in Hispania Tarraconensis include, on the Mediterranean coast, Rhode (mod. Rosas), Punta de l'Arenal, and New Carthage (mod. Cartagena); Atlantic sites include, especially, Gigia (mod. Gijón).[33] Most fish salteries, large and small, most likely operated independently of any state control.[34] The larger salteries had capacities well beyond what local needs would require, and, doubtless, exported a considerable amount of processed fish. The province of Lusitania (primarily modern Portugal), for example, has yielded the second largest Roman salting installation so far uncovered, and even now it is not yet fully excavated. The installations at Tróia (mod. Setúbal) in the first and second centuries AD had a salting capacity of over 600 cubic meters, and may have ultimately reached over 750 cubic meters.[35]

The largest Roman salting installation so far discovered in the Western Mediterranean was located at Lixus, on the Atlantic coast of Mauretania Tingitana (mod. Morocco). Its ten factories had a salting capacity of over 1,000 cubic meters.[36] Other North African salting installations of significant size include, in Morocco, Tahadart and Cotta, and in Africa Proconsularis, Neapolis (mod. Nabeul) and Sullecthum (mod. Salacta).[37] Although evidence of salting has been found on the Mediterranean coast of southern France, the largest installations in Roman Gaul were located on the Atlantic coast in the Bay of Douarnenez, at places like Plomarc'h.[38] And, finally, extensive excavations in the northern part of the Black Sea, especially in the Crimea at Chersonesos and along the Strait of Kerch at Tyritake and Myrmekion, have unearthed many well-preserved salting installations.[39] Unfortunately, these installations are little known outside of Eastern Europe. Indeed, though strongly hinted at in literary and epigraphic sources, salting installations in the Greek East have, generally, yet to be discovered.[40] Included among these are many sites located on the southern coast of the Black Sea, such as at Sinope and Byzantium, and along the Aegean coast of Turkey, at Clazomenae and Rhodes, to name only a few of them.

Although fish salting may have operated in the Black Sea as early as the seventh century BC, but certainly no later than the fifth century BC, archaeological excavations have yet to prove it. Finds from the fifth-century BC Punic Amphora Building in Corinth, however, do confirm references from fifth-century BC Attic comedy writers that indicate that salt-fish production and export were part of the economy of Punic colonies in the Western Mediterranean at that time. Punic fish sauce amphorae found in Corinth came either from North Africa or southern Spain. Some of them still contained rectangular bits of preserved fish, perhaps τετράγωνον. Additionally, finds of late fifth-century BC Punic salting installations at Las Redes, near Gades, substantiate an active salting industry in the Western Mediterranean at this early date.[41] Most Roman salteries date between the first century BC and the fourth century AD, with some operating into the sixth century AD.[42]

Literary sources can tell us how the ancients prepared fish by-products and can often indicate where they were produced, while archaeology, by revealing the physical remains of the installations themselves, can confirm these locations and disclose others. They cannot, by themselves, however, provide an understanding either of the physical and chemical processes the fish underwent to become the desired product, or of the nutritional and medicinal value that the ancients attributed to fish by-products. For this we must turn to modern food scientists and present-day manufacturing installations. Although certain products similar to ancient fish sauce are still being produced in parts of France, Greece, and elsewhere in areas that made up the Greco-Roman world, the most instructive comparative material can be found in Southeast Asia, in coastal areas of Vietnam, Thailand, and the Philippines.[43]

Recent studies of modern fish by-products, such as the salted herring and anchovies processed in Russian and north Atlantic salteries and the Southeast Asian fish sauces, including Vietnamese *nuoc-mam*, Thai *nam-pla*, and Filipino *patis*, indicate that present-day production methods, for the most part, parallel almost exactly those used in the Greco-Roman period. Biochemical and microbiological analyses of modern fish sauces tell us much about the Graeco-Roman examples described by ancient authors or, in some cases, uncovered by archaeologists. Whereas there are many modern methods for preserving fish by-products, ancient processes for producing fish sauce involved primarily autolysis, that is, a fermentation process of enzyme hydrolysis utilizing naturally occurring enzymes found in the digestive tract.[44]

Among variables that lead to different fish by-products are species of fish, type of salt, fish-to-salt ratio, length of processing, and minor ingredients. Of these, the species of fish in particular affects the product's nutritional value, as well as its taste, colour, and odour. Biochemical and microbiological studies have shown that fish sauce is composed of proteins in the form of amino acids, such as lysine, and of peptides, and contains numerous vitamins and minerals, such as vitamin B_{12}, sodium, calcium, magnesium, iron, manganese, and phosphorus.[45] The ancients, of course, did not know of vitamins and minerals and the like. They could only comment on the sauces themselves, noting physical characteristics and speculating on the presumed value to health born of observation and superstition.

The physical characteristics of ancient fish sauces can be conjectured from those of their modern counterparts. The taste of *patis* and *nuoc-mam*, for example, has been described as salty, with a distinct cheese-like taste; *nam-pla* has a "meaty" flavor. A recent series of studies, particularly in Japan, however, has identified in modern fish sauces significant quantities of monosodium glutamate (MSG), which, these scholars argue, imparted to the products a specific and identifiable taste, denoted *umami*, distinct from the standard four tastes of sweet, sour, salty, and bitter.[46] The color of the best Southeast Asian fish sauces varies between the clear, straw yellow to amber color of *patis* to the rather brown color of *nuoc-mam* and *nam-pla*. Scientific studies on modern fish by-products not only provide information important to our world but also produce significant data useful for understanding different aspects of the ancient world, such as health and nutrition. These investigations apparently sometimes work in reverse order as well. One recent study, for example, in attempting to reproduce the ancient *garum*, claims to have created more quickly a fish by-product that is even more nutritional than its modern counterparts.[47] If anything can be made of this, then ancient fish sauce might provide a practical contribution to the modern world.

3. *Commerce*

The same types of ancient literary evidence that supplied information on production of fish by-products also provide important data about their trade. These include histories, orations, medical treatises, geographies, encyclopedias, poetry, drama, gastronomic literature, agricultural manuals, private letters, and the like. Greek dramatists of the fifth and fourth centuries BC, for example, provide evidence of early trade between Greece, that is, Athens, cities of the Black Sea region, and Punic areas of the Western Mediterranean. This trade also finds a strong echo in later Greek and Roman authors.[48] Praise by writers in Rome or in Athens of fish by-products, probably often personally known to them through their availability in local markets, shows, or more often implies, that those products traveled in some fashion to get there. The prominence given to preserved fish products from Spain and the Black Sea by both Greek and Roman authors indicate that these were the two areas most active not only in producing but also in trading in fish by-products. Although literary sources provide us with valuable information on commerce from the point of view of the consumer living at the centre of importation, that is, in Athens and in Rome, they do little to illuminate the actual transportation of these goods or to identify individuals associated with their commerce. For that we must look to archaeological and epigraphic sources.

The artefact most important in providing information about commerce in salted fish products is one that began to receive proper attention only in the late 19th century. The amphora was the two-handle terracotta vessel used to transport food items long distances. In 1879 Heinrich Dressel established, albeit unintentionally, the first typology of Roman amphorae.[49] Basing his work on painted inscriptions (*tituli picti*) appearing on many vessels excavated on Monte Testaccio in Rome, he identified, among others, those amphorae that had held fish sauce or salt fish, and arranged them by shape. Since that time, and particularly in the last half of the twentieth century, other scholars have refined or added to this early typology or have created completely new ones, giving to them their own name or the name of the place where the vessels were discovered.[50] This has created a complex and confusing array of amphora shapes associated with fish by-products. One of the major questions yet to be answered is to what extent one can relate amphora shape to its contents and to its point of origin. So, for example, Dressel Forms 7-14, Pelichet 46, Beltrán I, Almagro 50, Camulodunum 186A, and Vindonissa 586, among many others, identify fish sauce amphorae from Spain, while Africana I and II may have carried fish by-products from North Africa.[51] Recent amphora studies have gone beyond shape to include not only chemical and fabric analysis of the clay used to make the vessels but also the *tituli picti* appearing on them.[52] This information along with the find spots of the vessels, such as shipwrecks whose cargoes contained amphorae, port cities, such as Rome, Ostia, and Pompeii, military camps, and the like, plus governmental, funerary and dedicatory inscriptions and papyri, have revealed a vast amount of information on trade in salted fish products. Four examples suffice to illustrate this point.

First, identification of the contents of amphorae has always perplexed scholars. Some vessels bear a painted inscription, or *titulus pictus*, that records the container's contents. Most amphorae, as extant, lack a *titulus* but their shape conforms to one or another type listed in various typologies. In this case, although we can reasonably conclude that the vessel once held a fish by-product, we do not know if the contents were fish sauce or *salsamentum*. Some amphorae, usually found among cargoes of ancient shipwrecks and lacking a *titulus*, still contain identifiable fish bones. Among the most prominent shipwrecks yielding amphorae with fish bones are the Sud Perduto II, Cap Béar III, Port-Vendres II, and Saint Gervais 3, from Spain, and the Grado from North Africa.[53] Since *garum, liquamen*, and *muria* were liquids, skeletal fish remains might represent either *allec* or *salsamentum*, but determining which one remains difficult. One recent study has begun to tackle this problem. Desse-Berset and Desse conclude that a container with many small whole fish, particularly clupeids, like sardines and anchovies, whose bones are disarticulated and mixed up, probably held *allec*. If the number of fish contained inside is relatively few and if the fish identified are larger than clupeids and the bones are generally intact and well preserved, the product was probably *salsamentum*.[54] This type of study is fairly recent however, so the question of criteria is far from settled.

Second, the painted inscription found on many – but not all – amphorae, has a standard and fairly consistent pattern, although not every label contains every item of information.[55] The kinds of information revealed include identification of the contents, along with any reference to their quality, and the ingredients used to make the sauce, such as the type of fish used. Following this the name of the owner of the vessel, the producer of the contents, or the person responsible for transporting the vessel frequently appear. Sometimes the recipient of the vessel might be listed. The *titulus* might also contain a number, of indeterminate meaning, that could be the vessel's weight, age of the product, or an indication of an imposed tax. One example comes from a one-handled vessel called the *urceus*, the vessel most often found in first-century AD Pompeii to have contained a fish sauce. The *titulus* reads: G(ari) F(los) SCOMBR(i)/ SCAURI/T(?) MAR/ L(uci) MARI PONICI.[56] The first line translates "the flower of garum, made from the mackerel." The next line reads "[a product] of Scaurus." In the third line appears an unknown symbol followed, after a space, by what appears to be an abbreviated name. The last line contains the name, in the genitive case, of "Lucius Marius Ponicus". The label has named the product (*garum*), declared its high quality ("the flower"), disclosed its ingredients (the mackerel), and identified the producer of the sauce (Scaurus). The meaning of the sigla is unknown; while MAR may refer to a manager of one of Scaurus' workshops, though the name does not appear elsewhere in Pompeii.[57] Ponicus may be the owner of the *urceus* or the shipper transporting the vessel. Therefore, by naming the contents, denoting its quality, specifying the product's ingredients, designating the producer, signifying perhaps its place of manufacture, and identifying the exporter, the *titulus* is at once a product label that includes information that would probably satisfy the United States Food and Drug Administration. It is as well a vehicle for product advertising.[58]

These labels are also excellent sources to learn about the individuals who participated in trade in fish products. Names appearing in *tituli* indicate that many, but not all, involved in the trade were freedmen. The *urceus* discussed above contained a product made by Aulus Umbricius Scaurus, a wealthy freeman living in Pompeii in the early to mid-first century AD.[59] This individual, to judge from numerous *urcei* bearing his *titulus*, dominated the fish sauce trade in Campania. Many *tituli* indicate that, in addition to products from his own shop, he utilised his freedmen to distribute his product from several other shops.[60] The inscription on his tomb indicates that his son rose to the highest magistracy in the city and had an equestrian statue erected in his honor in the forum at the expense of the city council. The unique mosaic floor installed in a secondary atrium of the house at Region VII. Ins. Occ. 12-16 identifies Scaurus' luxurious home.[61] This mosaic had the design of an *urceus* at each corner of the impluvium. On each mosaic *urceus* is a *titulus* identifying either *garum* or *liquamen*, products made and sold by Scaurus.

Third, Scaurus' *urceus*, carried by L. Marius Ponicus, was actually found not in Pompeii but at Fos-sur-mer at the mouth of the Rhône River in southern France. How it got there provides an important source for commerce in salted fish products. In recent decades underwater archaeology has expanded to include deep and shallow water finds of Greek and Roman ships wrecked for various reasons.[62] Most contained cargoes of amphorae not only of wine and oil but also of fish by-products. Study of the individual amphora provides important information of the kind described earlier. Plotting shipwrecks that contain fish sauce amphorae provides a graphic view of the usual sea routes followed by merchant ships. For example, a primary trade route between Spain and Italy, plotted by shipwrecks containing salt-fish amphorae, ran from Spain northward along the Mediterranean coast to the mouth of the Rhône River. From there ships headed east where the shipping lane split into two routes. One route went north of Corsica, the other ran between Corsica and Sardinia. From there ships could head to Rome, to the Bay of Naples, or elsewhere, including the Near East where Spanish salt fish amphorae have been found.[63]

At the mouth of the Rhône River, sea-going ships could offload their cargo onto riverboats that could head north into the heart of Europe. Plotting amphora finds along major rivers, such as the Rhône and Rhine Rivers, can also identify interior trade routes, by which fish by-products from Mediterranean salteries made their way to soldiers, government functionaries, and others with a taste for sea fish.[64] Fish sauce produced locally in northern

Europe also found its way into long-distance commerce. For example, recent finds in the interior of Belgium of the bones of small sea fish (mainly sprats and unidentified clupeids) have been interpreted as evidence of local transport and trade in fish sauce from the northeast coast, perhaps in the vicinity of Colijnsplaat.[65] Dedicatory inscriptions, found at Colijnsplaat in Germania Inferior and dating to the late second or early third century AD, show that fish sauce merchants, *negotiatores allecarii*, carried their products, whether local or Spanish, across the Channel to Britain.[66] Excavations in London, York, along Hadrian's Wall, and in many other places show that fish sauce from Spain and elsewhere traveled a great distance from the Mediterranean.[67]

And finally, epigraphic evidence also provides other important information about trade in fish by-products. Although, with one exception, we lack evidence for a specific price charged for a definable quantity of salt-fish or volume of fish sauce, we can ascertain the relative value of these products. Tariffs on fish by-products, such as those from Bacchias (*P.Wisc.* II.80) in Egypt and Palmyra (IGRR 3.1056.ii.35) in Syria, both of the early first century AD, or from Zarai (CIL 8.4508) in Africa Proconsularis, dating to AD 202,[68] indicate that most fish by-products were not expensive, regardless of what some literary sources might imply.[69] This is also borne out in Diocletian's *Edict of Maximum Prices* (AD 301), which specifies a highest price allowable for an amphora of fish sauce of two different qualities (III.6-7). Comparing these prices with maximum prices for other common items listed in the same document, such as honey and pork, prices for fish sauce compare relatively well. This is the strong implication as well from find spots of fish sauce containers in first-century AD Pompeii, where vessels have appeared in kitchens and gardens of houses both of the rich and of the poor alike.[70]

These examples, among many others that could be cited, suffice to give an idea of the wide range of sources now available to study the production and trade of Greco-Roman salt-fish products. I have also emphasised the individuals working today in various professions who are cooperating to discover, to analyze, and to interpret the evidence. Scholars studying these products from various angles have provided us with a far more complete understanding of them than was possible when we were restricted to literary sources alone. We now call upon historical, archaeological, epigraphic, papyrological, and art historical evidence. Studies on modern equivalents to Greco-Roman fish by-products provide comparative data that augment our knowledge of the ancient products. Scientists, including ichthyologists, biochemists, and microbiologists, have taken a more active role in assisting the efforts of ancient historians, literary critics, archaeologists, epigraphers, papyrologists, and art historians. Together they have amassed a wealth of information about a food product that played an important role in many areas of Greco-Roman society. Some of these sources, however, have hardly been tapped and much more evidence is yet to be discovered. If results accomplished in the thirty-eight years since the appearance of Ponsich and Tarradell's work are any indication of the future, the coming years will bring even more gains and exciting discoveries.

Notes

1 I wish to thank Pia Guldager Bilde, Director of the Danish National Research Foundation's Centre for Black Sea Studies, University of Aarhus, for inviting me to participate in the workshop on marine resources and trade in fish products in the Black Sea region in antiquity. I also wish to express my appreciation to Dr. Tønnes Bekker-Nielsen, of the University of Southern Denmark, for his hospitality and kindness during my stay in Esbjerg, and to the other participants in the workshop for providing an interesting and stimulating discussion.
2 Curtis 1991, 114, 118-119. The ancient Egyptians and Mesopotamians also processed fish, but it remains unclear to what extent the technology went beyond merely drying or smoking. That they salted fish seems likely, but the archaeological evidence is far from conclusive. See Curtis 2001, 173-175, 238-240 .
3 For example, Constantine Harmenopulus *Manuale legum sive Hexabiblios* 2.4.22 (n.d., but Byzantine period), Theophanes Nonus *De omnium particularium morborum curatione* 156, 158, 162 (10th century), Liutprand *Relatio de legatione Constantinopolitana* 20 (10th century), and Symeon Sethus *Syntagma de alimentorum facultatibus* passim (11th century). See Curtis 1991, 184-190.
4 Rabelais in Nock and Wilson 1931, 1, 930; Rondelet 1554-1555, 141; Belon 1555, Chapt. 25, 72. See also Curtis 1991, 186-187, 190; French 1986, 263; Chibnall 1975, 57-78; and Gudger 1924, 269-281.
5 Köhler 1832. That same year a brief article also appeared on the mackerel and ancient *garum*, Cuvier and Valenciennes 1832, 286-294.
6 Smidth 1875. First published in Danish in 1871, it was reprinted in English translation in 1875. C. Badham 1854 and Blümner 1869 are also rather superficial.
7 Eberl 1892. The last half of the 19th century also saw the appearance of Joachim Marquardt's volume on the private life of the Romans, which treats the topic of processed fish in only a few pages. He does, however, make some use of painted inscriptions (*tituli picti*) on amphorae. See Marquardt 1893, 2: 60-68.
8 Rhode 1890.
9 Zahn 1910, 841-849. The role of amphorae and their written inscriptions (*tituli picti*), as early as 1879, had already been recognized for their importance in studying ancient trade. See Dressel 1879, 36-112, 143-195. See also Remarck 1912. Blümner's updated volume on Roman private life, like Marquardt's before him, devotes little space to processed fish, and likewise adds little to Köhler's work. See Blümner 1911, 184-188.
10 Grimal and Monod 1952, 27-38; Jardin 1961, 70-96.
11 Ponsich and Tarradell 1965.
12 The Russian archaeologist, Viktor Gajdukevič, had conducted extensive excavations in the Crimea since the 1930s, but his work was largely unknown in the West until publication of Christo Danov's excellent 1962 Pauly-Wissowa article, in German, on the "Pontos Euxeinos", followed in 1971 by Gajdukevič's own *Das Bosporanische Reich*, a revised German translation of his earlier Russian work, in which he summarized the results of his excavations. Danov 1962, esp. 955-985; Gajdukevič 1971, 376-378.

13 Some scholars, influenced by the works of M.I. Finley, continue to downplay the role of salt-fish in the ancient economy, relegating them to little more than temporary hedges against periods of food shortage and famine. Cf. Gallant 1985; Finley 1999.
14 Paoli 1975, 91. In opposition to this attitude, see Curtis 1983, 232-240.
15 Cf., e.g., Étienne and Mayet 1998a, 147-165; Paterson 1998, 149-167; Garnsey 1998; Haley 1990, 72-78; D'Arms 1981.
16 Cf., e.g., Parkins 1997.
17 The fragments, with English translation, of Archestratos are most conveniently collected in Olson and Sens 2000, while Athenaios' work can be consulted in the Loeb edition of that author. Flower and Rosenbaum 1958 contains a translation of Apicius, which, while ascribed to the noted cook of the first century AD, dates to the fourth century AD.
18 Curtis 1991, 131-141; Drexhage 1993, 27-55.
19 Specific references can be found in Curtis 1991, 6-15.
20 E.g. *allec* vs. *hallex*, or ὁ γάρος vs. τὸ γάρον. See Curtis 1991, 7, n. 7 and 8, n. 9.
21 Curtis 1991, 7, n. 2.
22 Ps.-Rufius Festus *Breviarium*; Ps.-Gargilius Martialis 62. See Curtis 1991, App. I-4 and I-5, respectively.
23 For Paris Bibliothèque Ms. Lat. 11219, see Lestocquoy 1952, 185-186. For an English translation of this section of the *Geoponika*, see Curtis 1991, 12-13.
24 For strained fish sauce, see Apicius *de re Coq*. 7.6.14; *Geoponika* 20.46.2; CIL 4.7110.
25 For the text of Artemidoros, see Pack 1963, 72. Cf. Manilios *Astronom*. 5.670-674.
26 Xenokrates *ap*. Oribasios *Medical Collections* 2.58.133-152. For salt fish products in ancient medicine, see Curtis 1991, 27-37. While Xenokrates' work is unavailable in English translation, Galen's *De alimentorum facultatibus* (*On the* Properties of Foodstuffs) can be consulted in Powell 2003.
27 Voskresensky 1965, 117-128; Van Veen 1965, 227-250. Strabon is readily accessible in the Loeb edition.
28 Ponsich 1988. Since that time, four other important works covering the salteries of Baetica and other parts of Roman Spain have appeared: Étienne, Makaroun, and Mayet 1994; Lowe 1997; Lagóstena Barrios 2001; and Étienne and Mayet 2002.
29 Some vats at Vão, in Baetica, measured 1.50×1.03×1.85 m., while others at Caetobriga measured 4.00×3.70×2.00 m. See Curtis 1991, 53-54.
30 Sternberg 2000, 133-153. Similar finds have come from salting factories in Baetica at Quinta do Marim (Olhão) and in Hispania Tarraconensis at Troia. See Desse-Berset and Desse 2000, 84-92.
31 Curtis 1979, 5-23.
32 Curtis 1991, 50-51.
33 See most recently, Lagóstena Barrios 2001.
34 Curtis 1991, 148-152; Ørsted 1998, 13-35. For possible imperial participation in the production and trade in Spanish processed fish, see Liou and Marichal 1978, 131-135, No. 27, and Curtis 1991, 63.
35 Étienne and Mayet 2002, 96.
36 Ibid., 118; Ponsich 1988, 103-136.

37 Tahadart and Cotta: Ponsich 1988, 139-159; Nabeul: Sternberg 2000, 133-153; Sullecthum: Foucher 1970, 17-21. For North African salt-fish production, generally, see Ben Lazreg, Bonifay, Drine, and Trousset 1995, 103-142.
38 Curtis 1991, 74-76.
39 Curtis 1991, 118-126. See above, note 12.
40 Curtis 1991, 112-118, 129-131.
41 Curtis 1991, 47-48; Lagóstena Barrios 2001, 98-100.
42 Curtis 1991, 178.
43 Lopetcharat, Choi, Park, and Daeschel 2001, 65-88.
44 Curtis 1991, 15-22; Beddows 1985, 2: 1-39; Mackie, Hardy, and Hobbs 1971.
45 Curtis 1991, 22-24; Lopetcharat et al. 2001, 71-72.
46 Lopetcharat et al. 2001, 79-82. For *umami*, see especially Kawamura and Kare, eds. 1987, and Yamaguchi and Ninomiya 1998, 123-138.
47 Aquerreta, Astiasarán, and Bello 2001, 107-112.
48 Curtis 1991, 126-29.
49 Dressel 1879, 36-112, 143-195.
50 Cf., for example, Peacock and Williams 1986.
51 Curtis 1991, 39-44, 70, esp. Fig. 1, p. 42.
52 Peacock 1977, 261-278; Peacock and Williams 1986, 14-15.
53 Desse-Berset and Desse 2000, 75-82; Colls, Étienne, Lequément, Liou, and Mayet 1977; Auriemma 1997, 129-155; idem 2000, 27-51. Amphorae containing bones have also been found on shore on the island of Elba and at Olbia in Sardinia. See, especially, Bruschi and Wilkens 1996, 165-169.
54 Desse-Berset and Desse 2000, 91-95.
55 See, for example, Zevi 1966, 208-247; Curtis 1991, 197-200; Liou and Rodríguez Almeida 2000, 7-25, Étienne and Mayet 2002, 211-221. See also note 9, above.
56 Liou and Marichal 1978, 165, No. 69.
57 Ibid. Cf. Curtis 1991, 198-199.
58 Curtis 1984-1986, 209-228.
59 Curtis 1988b, 19-49.
60 Curtis 1988b, 28-33; Étienne and Mayet 1991, 187-194; Étienne and Mayet 1998b, 199-215.
61 *CIL* 10.1024; Curtis 1984b, 557-566.
62 See, for example, Colls et al. 1977, and Parker 1992, passim. For a Spain-Italy connection in the processed fish trade, see also Haley 1990, 72-78.
63 Curtis 1988a, 205-10; Curtis 1991, 143-144; Cotton, Lernau, and Goren 1996, 223-238; Lernau, Cotton, and Goren 1996, 35-41; Meijer 2002, 142-145.
64 Curtis 1991, 80-83; Martin-Kilcher 1990, 37-44, idem 1994 and idem 2003. Cf. also the first-century AD Spanish vessel of *garum scombri*, found in Mogontiacum (mod. Mainz), that bore a *titulus pictus* indicating that the recipient was the imperial legate P. Pomponius Secundus. See Ehmig 1996, 25-56.
65 Van Neer and Lentacker 1994, 53-62.
66 Curtis1984a, 147-158; Immerzeel 1990, 183-192.
67 Curtis 1991, 79-85; Carreras Monfort 2000, 141-149; Jones 1988, 126-131.
68 Some *tituli picti* may have included indication of an export tax levied at the port of embarkation. See Frank 1936, 87-90. But cf. Colls et al. 1977, 95-98; Ehmig 1995, 120-125.
69 For expensive fish sauce, see Pliny *HN* 31.94; Seneca *Letters* 95.25; Manilios *Astron.* 5.671; Martial 13.103.
70 Curtis 1991, 170-175.

The Archaeological Evidence for Fish Processing in the Western Mediterranean

Athena Trakadas

1. *Introduction*

The evidence for fish processing in the western Mediterranean in antiquity is found in diverse literary and archaeological sources. Although the textual evidence for the industry in this region is more extensive than in any other, it is the archaeological evidence comprised of transport amphorae, coins, artefacts such as fishhooks or faunal remains, and the actual fish-processing sites themselves that offer a relatively clear view of the facets and extent of the industry. In particular, these sites clarify and further illuminate the processing of fish as *salsamentum* (τάριχος) and sauce (*garum, liquamen, muria, allec,* and *lymphatum*) mentioned or implied in texts.[1]

2. *Origins*

The indigenous populations of the western Mediterranean region undoubtedly practised fishing as a means of sustenance,[2] but the techniques of processing fish into *salsamentum* and sauce for later consumption were most probably introduced by peoples from elsewhere in the Mediterranean basin. It has been proposed that fish processing in the region arrived with the first colonisers, and so had Phoenico-Punic origins.[3] R. Étienne proposes another theory, suggesting that it was possible that the Phocaeans first introduced fish processing to the Punic colonists of the southern Iberian Peninsula, after arriving in the region from Asia Minor, where they had practised fish-preservation techniques since the seventh century BC.[4]

A majority of the earliest Phoenician settlements in the southern Iberian Peninsula, such as *Abdera, Sexi,* Chorreras, Toscanos, *Malaca,* and Guadalhorce (dating to the middle of the eighth century BC), were located on the southern Mediterranean coast of the peninsula,[5] but only the faunal remains of fish have been excavated at these sites, as at other coastal Phoenician settlements in Portugal and Morocco.[6] The coins of the Phoenician settlements of *Sexi* and *Abdera,* like *Gades,* depict fish (believed to be tunny) (Fig. 1),[7] and it is tempting to think that the first Phoenician colonies in the western Mediterranean *initially* focused upon the rich resources of the sea, much like the Greek colonies in the Black Sea region.[8]

However, the earliest archaeological evidence of fish processing in the western Mediterranean has been discovered in the subsequent Punic layers of one of the main Phoenician settlements, *Gades* (Cadiz). Four Punic fish-salting installations have been identified: Plaza de Asdrúbal, Avda. De Andalucía, Avdas. García de Sola y de Portugal, and Las Redes. The first three sites all have the implications of processing: fish bones, other organic debris, and Mañá A4 and Mañá D-type amphorae containing fish remains (Fig 2).[9] The site at Las Redes is the best preserved and still possesses the remains of processing facilities. In a small building at the site, there is a room for the cleaning and preparation of fish (with a paved and sloping floor), a fermentation room (with organic debris and possibly a hearth), possibly a room for macerating fish, a storage area for amphorae, and a room with fishing accoutrements such as fish hooks and line sinkers. Las Redes and the other Cadiz sites began to operate in late fifth century BC, with the height of activity occurring ca 430-325 BC; eventually the sites ceased operation ca 200 BC, around the time that the Roman organization of the province of *Baetica* began in earnest, after 197 BC.[10]

This archaeological evidence for fish processing in Cadiz is also linked to evidence of the exportation of salted-fish products to the eastern Mediterranean. The Mañá A4 and Mañá D-type amphorae found at Las Redes have also been discovered in central Greece.[11] Excavated in the so-called "Punic Amphora Building" adjacent to the agora at Corinth, the amphorae (dated to the middle of the fifth century BC) contained fish bones of sea bream and tunny.[12] Evidence of this trade in the fifth and fourth centuries BC is also corroborated by the Attic comedic writers Eupolis, Nikostratos, and Antiphanes, who specifically mention salted fish imported into Greece from *Gades*.[13]

Fig. 1. A coin from Abdera, on the southern Spanish coast, which depicts fish (tunny?) as columns of a temple (after Ponsich and Tarradell, 1965, Pl. XXIV).

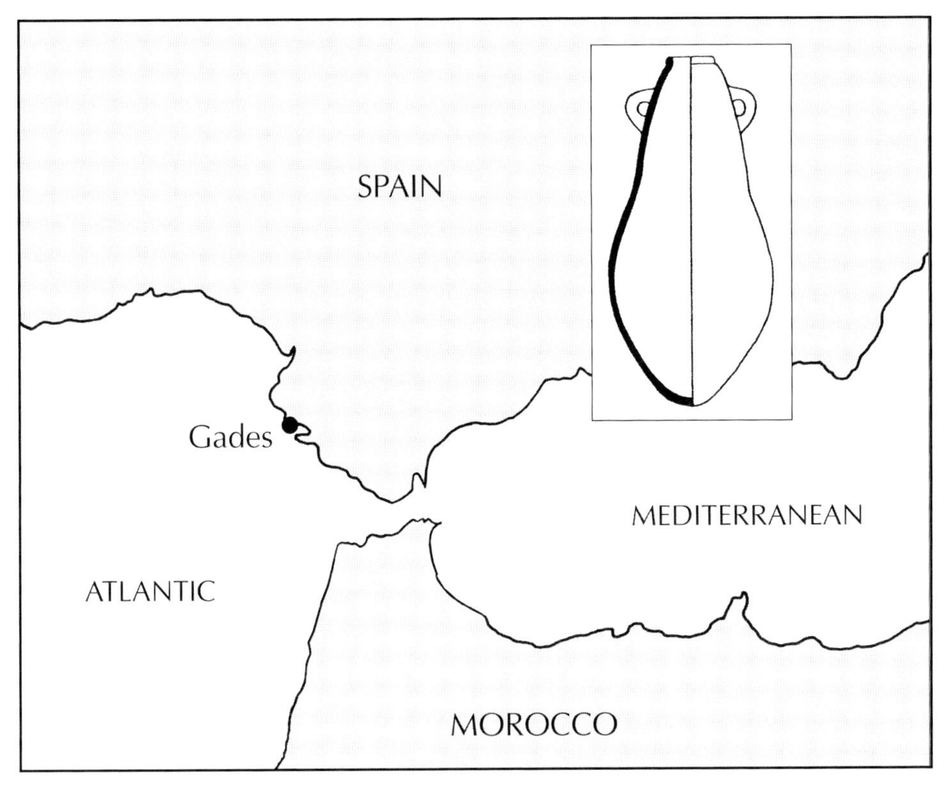

Fig. 2. The sites at Gades (Cadiz) have revealed evidence of fish processing starting in the late fifth century BC. Archaeological evidence includes Máña A4-type amphorae (after Muñoz Vicente, et al. 1988, fig. 9).

The site recently discovered at Las Redes reveals therefore that fish processing in the western Mediterranean was initially Punic in origin, which supports in part earlier proposed theories. That no other contemporary sites have been identified might be due to the fact that the numerous, later Roman fish-processing installations throughout southern Spain, Portugal, and northern Morocco probably removed any evidence of earlier installations, since they were often built on top of Phoenico-Punic sites;[14] also, archaeological vestiges could have been heavily damaged during the Second Punic War.[15]

3. *Fish-processing sites in Spain, Portugal, and Morocco*

After the Punic sites at Cadiz ceased operating ca 200 BC, there is a lacuna of over a century in the archaeological record throughout the region, no doubt due to the extensive geo-political transformations of Roman provincialisation. By the first century BC, however, fish processing in the region re-emerges as a technique practised by the Roman residents of the coastal zones, and a much more detailed picture of the industry is visible (Fig. 3).

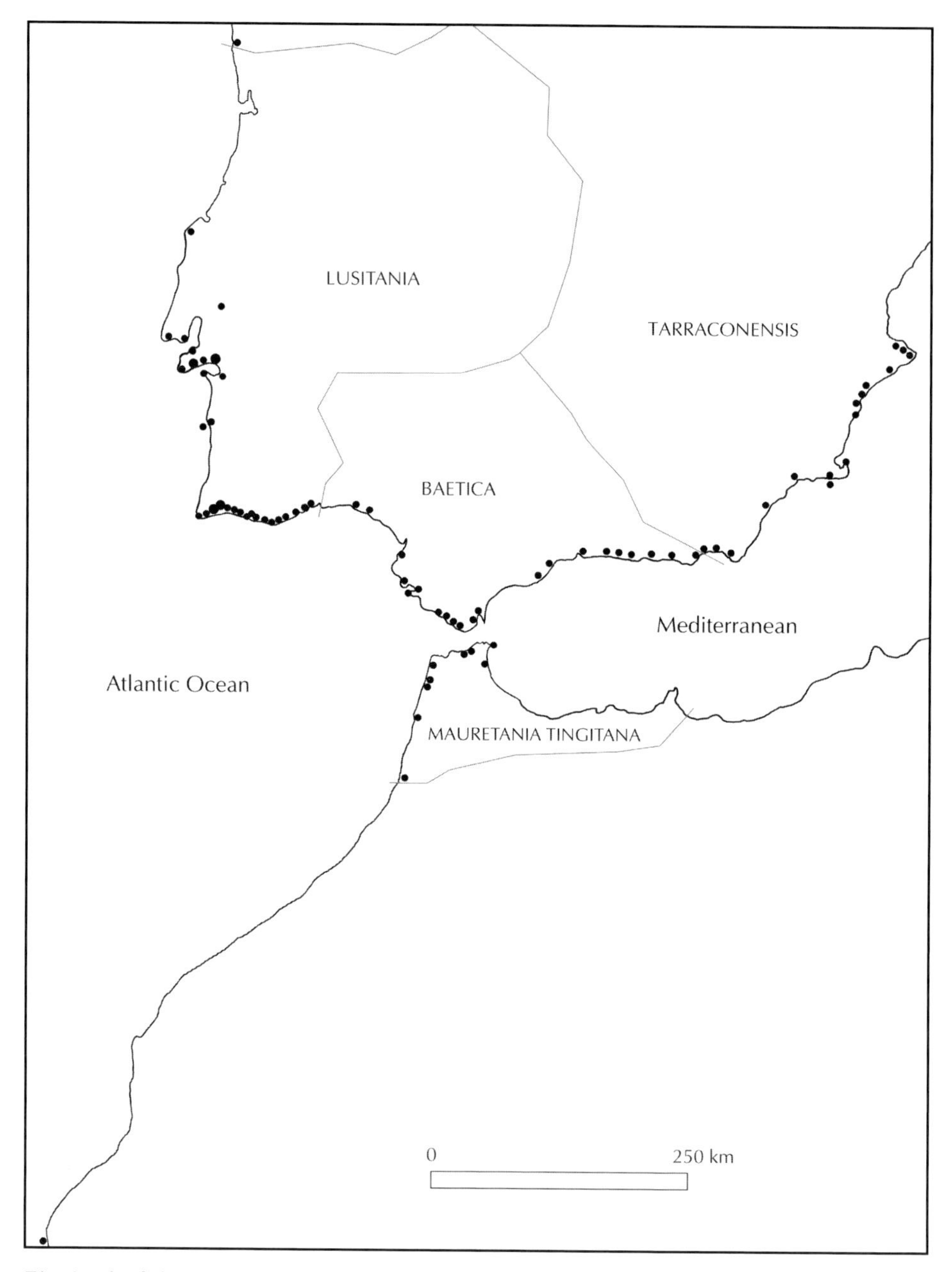

Fig. 3. The fish-processing sites were distributed throughout the Roman provinces of the western Mediterranean.

3.1 Spain

After the Second Punic War, and the fall of Numantia in 133 BC, the occupation and domination of the southern Iberian Peninsula by the Romans began

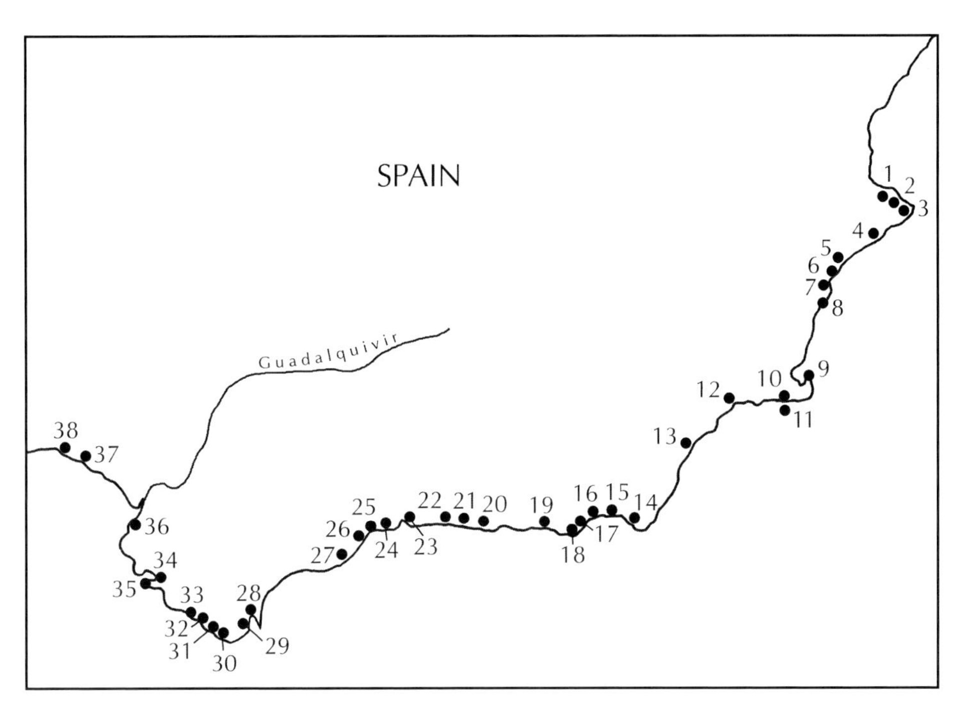

Fig. 4. The fish-processing sites in Baetica and Tarraconensis. (For key to site numbers, see p. 76)

in earnest.[16] By the Augustan period, several sites for the processing of fish began to develop along the Mediterranean and Atlantic coastlines of the peninsula, and the exportation of their products at this time is demonstrated by finds of southern Spanish fish-sauce amphorae as cargo of the mid-first century BC "Le Titan" shipwreck (found off southern France),[17] and in the Augustan levels at La Longarina, Ostia.[18] Additionally, fish products from Spain, specifically *garum,* are also documented by contemporary literary sources such as Horace (*Sat*. 2.8.46) and also Strabon (3.2.6).[19]

The Roman fish-processing sites in southern Spain were situated along the southern coasts of the provinces *Baetica* and *Tarraconensis,* spanning the rich waters of the western Mediterranean and eastern Atlantic. The locations of these sites were, and in many cases still are, ideally sited near the off-shore migratory routes and spawning grounds of many different types of fish, including tunny, mackerel, mullet, and eels.[20] During the period of Roman presence in southern Spain, numerous fish-processing installations existed; thirty-eight sites have been identified and are included here in this present survey (Fig. 4). From east to west, these include Denia (*Dianium*), Punta de Castell, Punta de l'Arenal (or Jávea), Calpe, Campello, Tossal de Manises (*Lucentum*), the island of Tabarca, Santa Pola (*Portus Illicitanus*), Mar Menor, Cartagena (*Carthago Nova*), *Scombraria*, Mazarrón, Villaricos (*Baria*), Torre García, Almería (*Portos Magnos*), Ribera de la Algaida, Roquetas, Guardias

Viejas, Adra (*Abdera*), Almuñúncar (*Sexi*), Torrox, Cerro del Mar (*Maenuba*), Málaga (*Malaca*), Guadalhorce, Torremolinos, Fuengirola, San Pedro de Alcántara (*Silniana*), El Rocadillo (*Carteia*), Algeciras (*Julia Traducta*), Villavieja (*Mellaria*), Bolonia (*Baelo*), Barbate (*Baesippo*), Trafalgar, Puerto Real, Cadiz (*Gades*), Sanlúcar de Barrameda, Cerro del Trigo, and Huelva (*Onuba*).[21]

Some of these sites in southern Spain have been more completely excavated than others, and therefore it is possible to assign only a very general chronology for the entire group. It is clear that some fish-processing activity did begin in the first century BC at Bolonia (*Baelo*), as the presence of one salting vat confirms,[22] and also at the small installations at Punta de l'Arenal,[23] Sanlúcar de Barrameda, El Rocadillo (*Carteia*), and Villavieja.[24] Most of the other fish-processing sites in *Baetica* and *Tarraconensis* began to operate in the first and second centuries AD, and more is known about this industry here during this period than in any other part of the Empire. Most sites stopped functioning completely in the third century AD, while others severely curtailed their production. Some even witnessed a later renewal in activity after the third century, and a few show signs of continuous but reduced operations until the sixth century.[25]

The sites in southern Spain vary in size, from only a few isolated salting vats, or *cetariae*, to entire complexes of these associated with small settlements, villas, or towns. Despite the difference in size, however, the basic features of these installations still reveal much typographical, constructional and functional homogeneity, visible in other sites in Portugal and Morocco. Throughout southern Spain, *cetariae* were constructed sunken into the ground and vary in size, although they are consistently rectangular or square in shape.

Within southern Spain, however, some constructional variations do occur in the fish-processing sites, perhaps based somewhat upon the slightly differing topography throughout the region. The factories along the coast in Alicante, between Santa Pola and Punta de l'Arenal, are frequently located on rocky promontories near the sea, with *cetariae* cut into the rock.[26] Uniquely, these sites also include fishponds (*piscinae* or *vivaria*), also cut into the rock.[27] Some of these ponds could be rather large, as at Punta de l'Arenal, where the so-called "Baños de la Reyna" measures 28×7 m and is 4 m deep (Fig. 5).[28] From the ponds, rock-cut channels led to the sea, likely functioning as feeder conduits, supplying fresh seawater into the tanks during the high tides.[29] As at other locations throughout the Mediterranean, these fishponds were probably used for keeping fish alive, prior to consumption, processing or live transhipment.[30] Strabon (3.2.7) notes that live murry caught in Spain were sent to Rome and if this did occur, the fishponds specific to the Alicante region perhaps played a role in this trade.

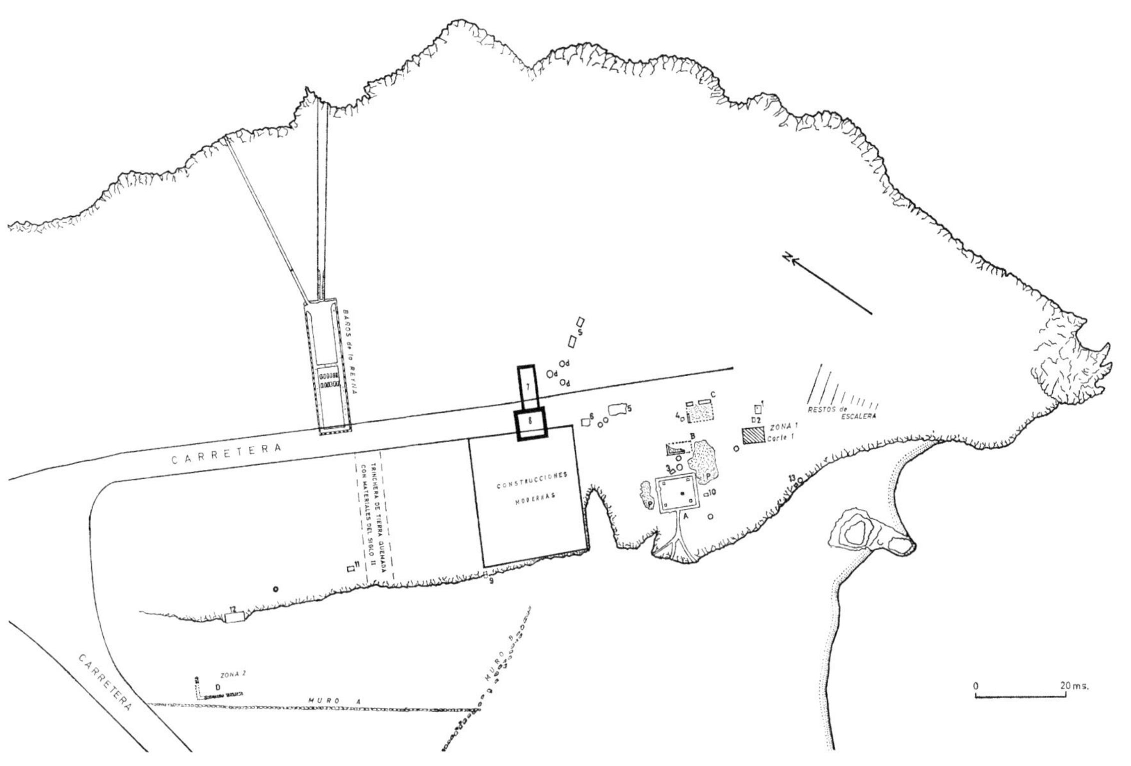

Fig. 5. The "Baños de la Reyna" at the fish-processing site of Punta de l'Arenal, southern Spain (after Martin and Serres 1970, fig. 2).

3.1.1 Baelo

Although villas were associated with some of the *cetariae* at Calpe, San Pedro de Alcántara, and Punta de l'Arenal,[31] almost all fish-processing sites in southern Spain were located some distance away from – or outside the walls of – major permanent settlements, most likely due to the strong, putrid smell incurred from the fermentation process. However, at *Baelo* (modern Bolonia), the fish-processing installations seem to have been located inside the walls of the city (Fig. 6).[32] This situation is unique in the western Mediterranean, but is reminiscent of the site of Tyritake on the Black Sea, located along the Strait of Kerch, where a large number of factories were situated in the southeastern part of the walled city (see p. 141-148).[33]

Baelo was the largest Roman fish-processing site in *Baetica* and *Terraconensis*. Located on the Atlantic coast of Spain at the western entrance to the Straits of Gibraltar, it was a port city that faces south, situated in a valley formed by a break in the coastal mountain range. Two streams, Arroyo de las Villas and Arroyo del Pulido, run through the small valley to the sea and are adjacent to *Baelo*.[34] Excavations have revealed that fish processing first began when

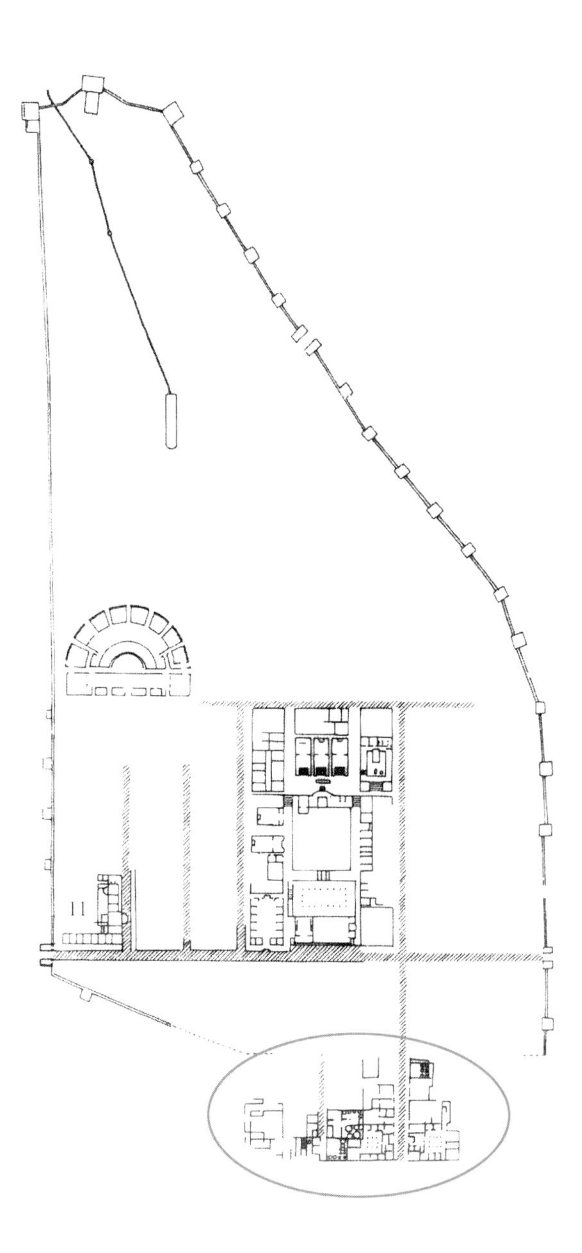

Fig. 6. The walled city of Baelo, with the fish-processing complexes in its southern sector (after Pelletier 1988, fig. 2).

the city was established in the first century BC.[35] As such, *Baelo* is one of the first sites in Spain to process fish (after Las Redes), and this is confirmed by Strabon (3.1.8), who mentions the *garum* industry of the city. A fish, thought to be a tunny, also appears on the reverse of the coins of *Baelo*, perhaps implicating the importance of fishing or the fish-processing industry to the city.[36] The fish-processing industry at *Baelo* declined at the end of the third century AD, but continued to operate until the fifth century or later.[37]

Two different areas in *Baelo* were utilised for fish processing: one is a group of small salting vats located to the south and outside the city itself, down along

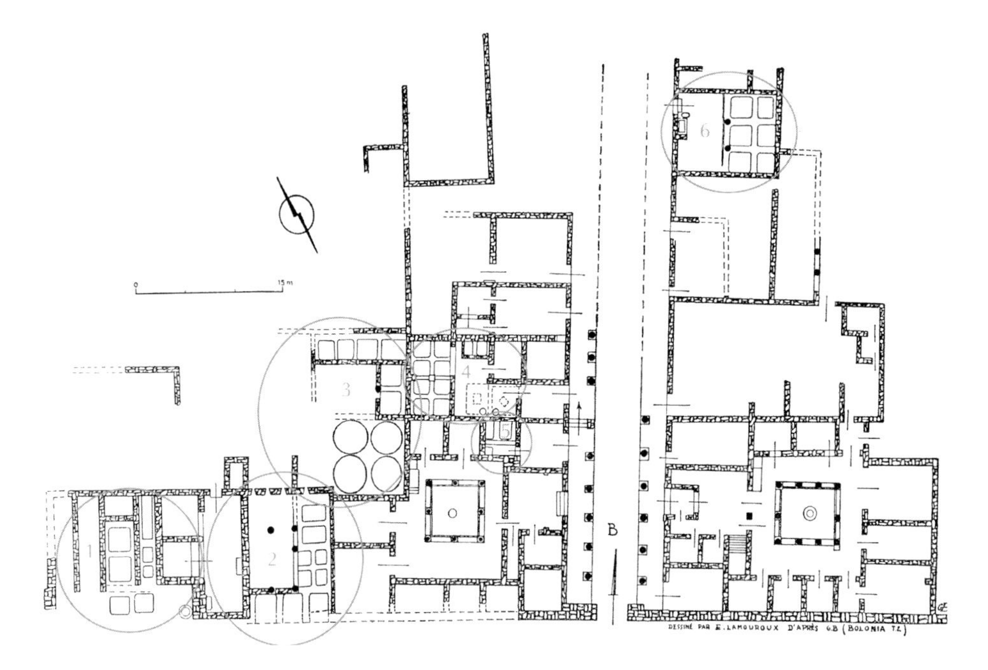

Fig. 7. The six fish-processing installations of Baelo (after Ponsich and Tarradell 1965, fig. 53).

Fig. 8. The four large circular salting vats at Baelo. Note the extant columns (photo: A. Trakadas).

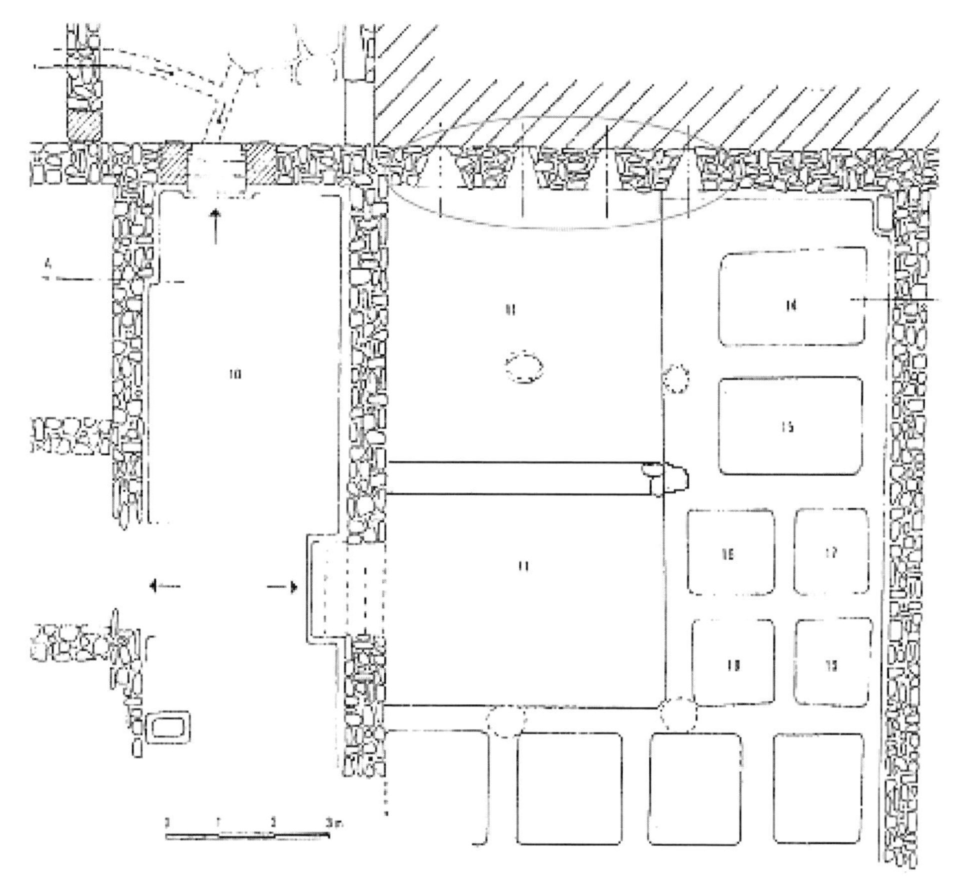

Fig. 9. The four windows in the wall of one of the complexes at Baelo (after Ponsich 1976, fig. 1).

the shore;[38] the second is a group of six complexes in the southern part of the city that faces the beach (Fig. 7). Of these latter installations, three had direct access to the beach, and three directly opened onto the *decumanus*. Nearby and adjacent to these installations are two peristyle houses.[39]

The installations in the city consist of differing numbers and arrangements of salting vats of varying size sunken into the ground; some are rectangular, some are square, and four rather large examples are circular (Fig. 8). It has been suggested that the larger rectangular basins, measuring ca 2×3 m and 1.6 m in depth, were probably used for the salting of fish to form *salsamenta*, and the smaller ones were used for making fish sauces such as *garum*.[40] The circular basins, the largest of which measures over 3 m across and 2.5 m deep, could also have been used for *garum*; M. Ponsich suggests they could indicate evidence of processing whale meat.[41] As Robert Curtis has pointed out, however, such circular vats could have also served for making *garum* from fish, as the shape would have facilitated stirring, necessary for an evenly autolysed mixture.[42]

Within each of the six complexes at *Baelo*, vats were located near a central "preparation" area, where the fish were probably cleaned and made ready for processing. One of these preparation rooms has a slightly sloping floor that inclines toward a sump, which probably was designed to assist in cleaning the facility by collecting organic refuse.[43] Water was carried by underground channels to the installations, and was used to help clean the facilities.[44] The installations in the city were covered with roofs and enclosed, most likely to prevent the unwanted rapid evaporation of the fish sauces brought on by direct sunlight, but the remains of four large windows in the wall of one installation support the theory that ventilation was desirable to the process (Fig. 9).[45] In all, the facilities at *Baelo* constitute a processing output of well over 220 cubic metres at any one time, an amount that undoubtedly exceeded local consumption needs.[46]

3.2 *Portugal*

The province of *Lusitania* was established when *Baetica* was reduced in 26-25 BC, and constituted what is now southern Portugal and a small portion of central Spain, from the Douro River south and from the Guadiana River west.[47] Remains of ancient fish-processing sites in Portugal do not pre-date the Roman period, contrary to earlier belief.[48] No sites have been identified that functioned during the Republican period, but Strabon (3.2.6) notes that fish processing occurred along the Algarve coast, implying that facilities were established by the first century BC. As in Spain, a few sites were operating during the latter part of the first century BC, but the major expansion of the industry occurred during the following two centuries.

Like southern Spain, the Atlantic coastal waters of *Lusitania* were – and are – rich in tunny and other pelagics, as well as shellfish. That fish were an important part of the livelihood of the region might also be demonstrated by the appearance of fish on the coins of several towns in the province: *Baesuris* and *Ossonoba* on the Algarve coast, and the inland river ports of *Salacia* and *Myrtilis*.[49] The Algarve and the Sado Estuary were the two main areas of fish processing exploited during the Roman period, but sites extend from the Guadiana River (the eastern border to *Baetica*) to the Douro River on the Atlantic coast.[50] Forty sites have been identified and are included in this present survey (Fig. 10). They are, from east to west: Quinta do Lago, Quinta do Muro, Cacela, Tavira (*Balsa*), Alfanxia, Olhão, Faro (*Ossonoba*), Loulé Velho, Quarteira, Cerro da Vila, Armação de Pera, Ferragudo, Portimões, Boca do Rio, Mexilhoeira Grande, Vau, Paul, Senhora da Luz, Burgau, Salema, Ilheu de Baleeira, Ilha do Pessegueiro (*Poetanion*), Sines, Tróia, Alcácer do Sal (*Salacia*), Santa Catharina, Senhora da Graca, Pedra Furada, Cachofarra, Setúbal (*Caetobriga*), Comenda, Rasca, Creiro, Alfarim, Casilhas, Lisbon (*Olisipo*), Guincho, Garrocheira, Atouguia, and Praia de Angeiras.[51]

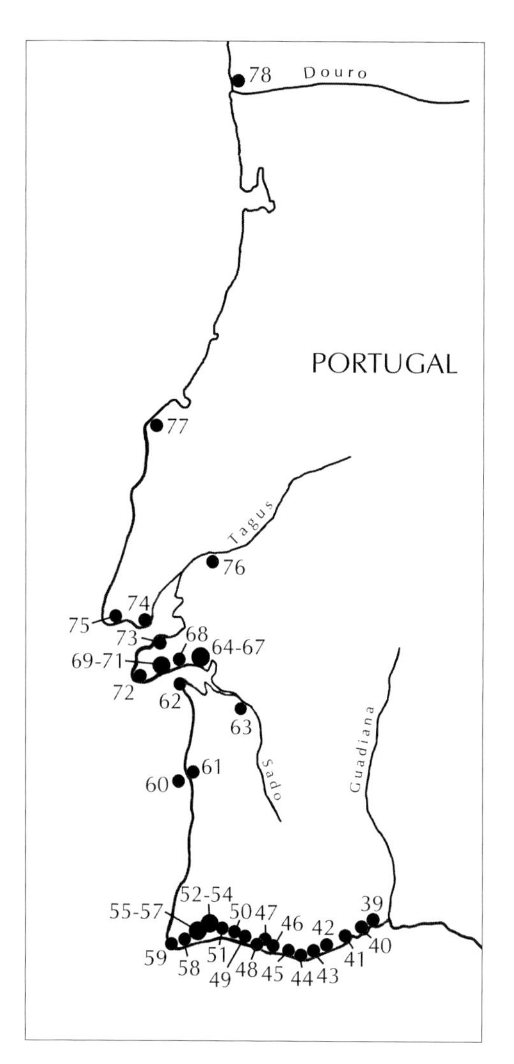

Fig. 10. The fish-processing sites in Lusitania. (For key to site numbers, see p. 77)

A majority of the sites in Portugal were noted as early as the late nineteenth century. Some installations were cursorily excavated at this time, but it is difficult to ascertain much detailed information from early reports due to methods of recording and confusion in stratigraphy.[52] Excavations in the past century, however, have led to a clearer picture of these sites and have assisted in establishing their general chronology. Two installations, the large site of Tróia and small site of Casilhas, both began to function at the end of the first century BC. The remainder of the fish-processing installations in Portugal, as in southern Spain, began operating mainly in the first century AD, and many of these continued to function until the beginning of the fifth century AD.[53] At some fish-processing installations, a portion of the *cetariae* went out of operation in the third century; at Setúbal, however, the bottoms of some

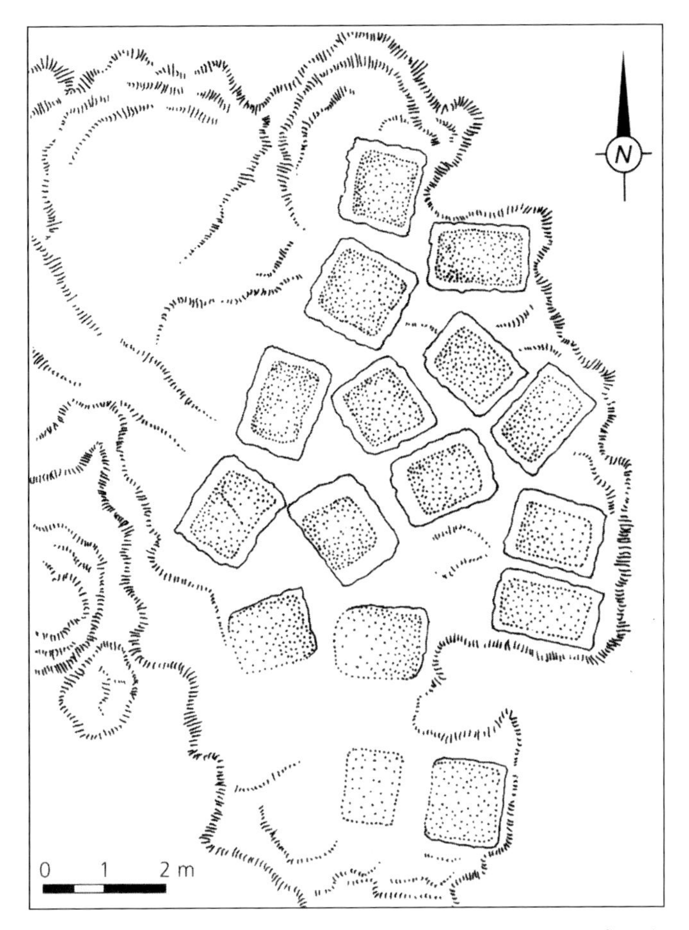

Fig. 11. The cetariae *cut into rock at Praia de Angeiras (after Gil Mantas 1999, fig. 4).*

cetariae were re-constructed, suggesting that part of the complex operated not only in the early Empire, but also again in the fourth and fifth centuries AD, after a period of disuse.[54] At Tróia, some installations continued to operate until the sixth century AD.[55]

Despite the fragmentary preservation of many small installations and scattered *cetariae*, several sites clearly reveal the extent of fish processing in Portugal. At Boca do Rio, in the Algarve, the remains of salting vats are preserved, although a large portion of the nearby settlement has been built over. The concentrations and number of *cetariae*, as well as the rich mosaics still extant from nearby residences, suggest that this was probably a large processing site that sold its products.[56] On Ilha do Pesseguiero, off the Atlantic coast near Sines, a fish-processing installation consisting of two complexes of 18 *cetariae* with storerooms has been excavated. The *cetariae* were dug into the bedrock, in a construction similar to that at Praia de Angeiras and reminiscent of the sites in the Alicante region in southeastern Spain (Fig. 11).[57]

As the majority in *Baetica* and *Tarraconensis*, fish-processing sites in *Lusitania* were located away from substantial urban settlements, and none were situated inside city walls, as were the installations at *Baelo*. In several locations, however, villas were located close by complexes or associated with scattered *cetariae*; this is mainly the case along the Algarve coast at Boca do Rio, Mexilhoeira Grande, Ferragudo, Cerro da Vila, Quarteira, Olhão, Paul, and Caecela, but also at the isolated site of Praia de Angeiras on the north Atlantic coast.[58] In some cases, the installations probably represent production for local consumption of the residents and dependents of the villas, while others that are more extensive constituted part of industrial annexes for the production of marketed goods.[59]

3.2.1 Tróia

The most extensive and largest fish-processing site in *Lusitania* is Tróia, located on a promontory that separates the mouth of the Sado River and the Atlantic in central Portugal. This promontory guards the entrance to the marine-rich Sado Estuary, but the sandy environment on which the site is located probably prevented any agricultural sustenance. It is therefore assumed that fishing and fish processing were the primary forms of activities in antiquity, and the extensiveness of the installations at the site would appear to confirm this.[60]

Tróia was one of the first sites to operate in *Lusitania*, with some processing installations operating in the late first century BC, but most initiating production by the middle of the first century AD. A substantial decrease in operations and production occurred in the second century; however, by the fourth century, a number of installations were modified for re-use or built over with other edifices such as chapels, or were used as cemeteries. In limited areas at the site, a certain level of fish-processing activity appears to have continued uninterrupted until the end of the fifth century or beginning of the sixth century.[61]

The fish-processing installations at Tróia extend for over 4 km along the western shoreline of the Sado River. The installations mainly consist of small units of salting vats spread along the length of the shoreline (much like across the river at Setúbal), with the greatest concentration of *cetariae* extending for almost 1 km (Fig. 12).[62] Fifty-two "units" of production have been estimated, and their individual plans are generally similar to the installations at *Baelo* and in Morocco, such as Cotta and *Lixus*.[63] The rectangular *cetariae* of Tróia differed in size and capacity, as at other sites; possibly this difference reflects various types or strengths of fish sauces. The smaller vats could possibly represent the more concentrated and therefore more expensive types of *garum*, while the larger vats, measuring ca 3×4 m, could represent cheaper types.[64]

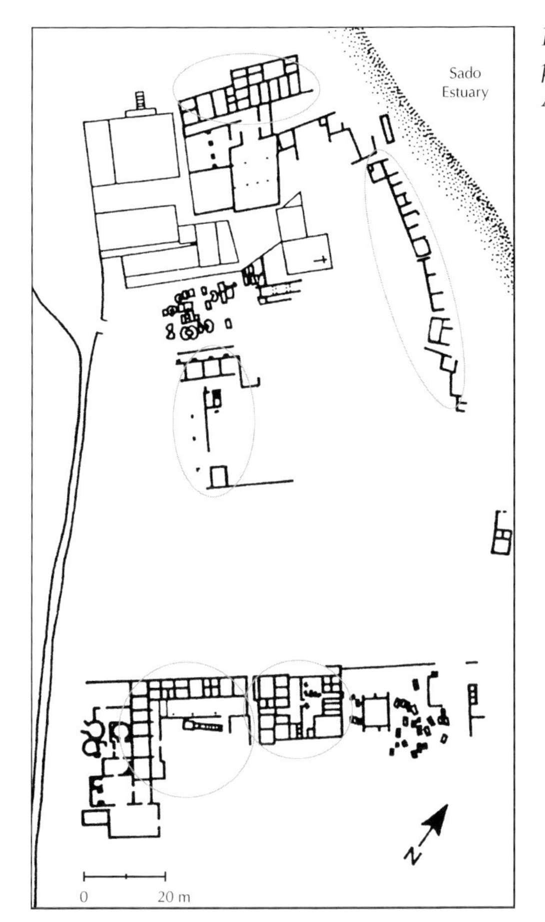

Fig. 12. The main concentration of fish-processing complexes at Tróia (after de Alarcão 1988b, fig. 130).

Some *cetariae* at Tróia were located in long rows running parallel to the shoreline. Other installations were situated in complex-like buildings slightly inland. The largest of these latter installations are the so-called "Factories I and II", which were adjoining complexes in the centre of the peninsula. These two factories also clearly display the chronology of the site itself (Fig. 13). Initially, Factory I covered a large area, with roofed vats encircling a large, open courtyard with a central well and cistern. During this first phase, which began in the middle of the first century AD, there were 19 extant vats that varied in size, the largest of which measuring 3.75×4.0 m and 2.4 m deep and the smallest measuring 3.6×1.5 m and 1.93 m deep. The volume of the extant vats was 465 m^3, but the entire complex is estimated to have been ca. 700 m^3.[65] Connected to the first installation, but similar in layout, was Factory II, which was smaller than the first with only 11 extant vats of almost uniform size. The total volume of this factory was 141 m^3. Also adjoining this complex were storage facilities for amphorae.[66]

These factories were abandoned at the end of the second century AD, but re-occupied and modified at the beginning of the fourth century AD. At this

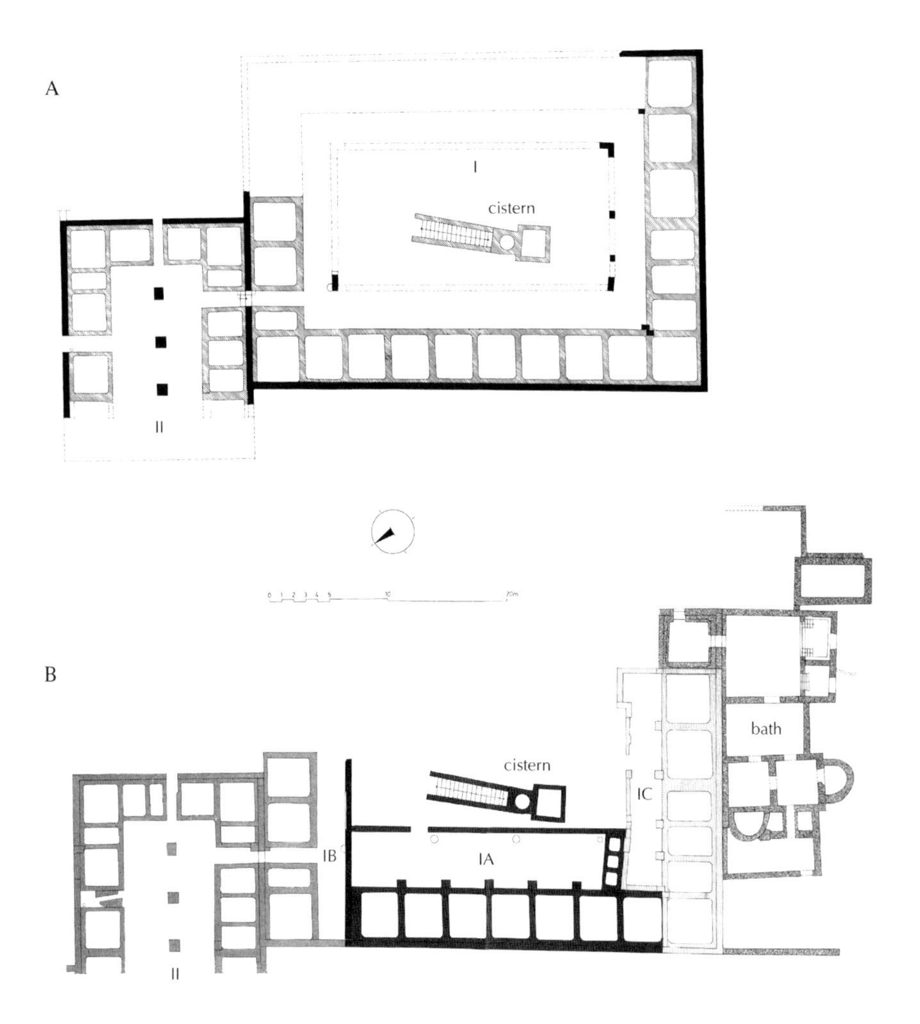

Fig. 13. The first and second phases of "Factories I and II" at Tróia. During the third phase, the cetariae *of Factories IA, IB, and IC were further subdivided (after Étienne, et al. 1994, figs. 55-56).*

time, Factory I was divided into three smaller units, called "Factories IA, IB, and IC," and several of the *cetariae* were also subdivided. At the same time, a bath was also built adjoining Factory IC, and one of the original salting vats was re-used as a washbasin for this building.[67] In the third phase of use, more of the vats were subdivided, creating smaller vats and smaller production outputs. Finally, the factories ceased production at the end of the fifth century.[68]

The preparation of fish at the factories at Tróia would have taken place in the open space in front of the *cetariae*, and in some instances this area was

Fig. 14. A visualisation of parts of Factories IC and IA with roofs, with the adjoining bath complex at the rear (after Étienne, et al. 1994, fig. 48).

usually paved with the same waterproof material as the vats themselves. In some instances, like at *Baelo*, the floors of these preparation rooms sloped slightly, draining towards a sump to collect the organic refuse from the cleaning process. Evidence of pillars suggests that some complexes, like Factories I and II, were covered with a roof, and openings for ventilation were no doubt present in surrounding walls (Fig. 14).[69] Fresh water was supplied to the complexes of Tróia by means of a system of cisterns and wells distributed throughout the site.[70]

The level of industry that took place at Tróia probably attracted the development of a semi-urban community that was directly involved in fishing and fish processing or in other related services.[71] The large population present at Tróia lived in houses with rich mosaics and murals that were situated adjacent to and amongst the fish-processing installations themselves. These houses, the presence of administrative buildings, a forum, as well as the number of vats at the site, suggest year-round fish-salting production. With extensive kilns also located in the region, the site most likely was a major commercial *vicus*, with a production output that far exceeded local requirements.[72]

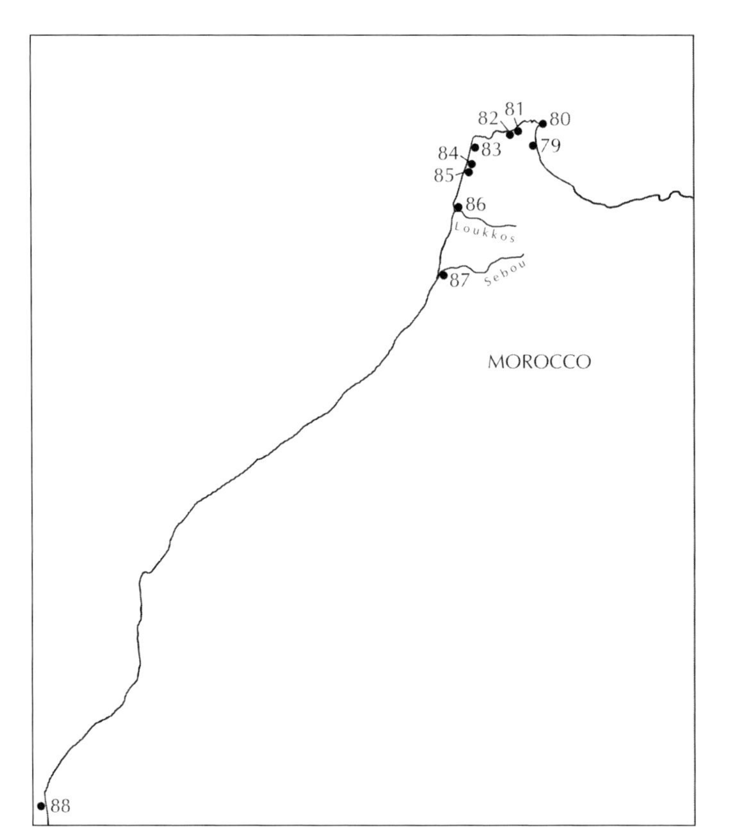

Fig. 15. The fish-processing sites in Mauretania Tingitana (For key to site numbers, see p. 78).

3.3 Morocco

Although Strabon (3.2.7) mentions the presence of tunny just outside the Straits of Gibraltar along the coast of North Africa, there exists a lacuna in the literary record regarding fishing and fish processing in Morocco in antiquity. Even though the residents of Phoenician and Punico-Mauretanian settlements along the Atlantic coast of northern Morocco certainly exploited the rich marine resources, the archaeological evidence for the processing of fish coincides with the Roman influence and colonization in the region, in the first century BC.[73] When northern Morocco was annexed as *Mauretania Tingitana* in 43 AD, the Roman province extended south to the Oued Bouregreg on the Atlantic coast, but included the distant Îles Purpuraires at Essouaira. It is during this century that fish-processing sites began to develop fully in the province.[74]

The fish-processing sites in *Mauretania Tingitana* are not as numerous as those across the Straits of Gibraltar in *Tarraconensis*, *Baetica* and *Lusitania*, but they are better documented. These sites stretch from the Mediterranean to the Atlantic coasts, adjacent to waters that were – and still are – rich with tunny, mackerel, sardines, and eels, as well as shellfish. Ten Roman-period fish-processing sites have been identified and are included in this present

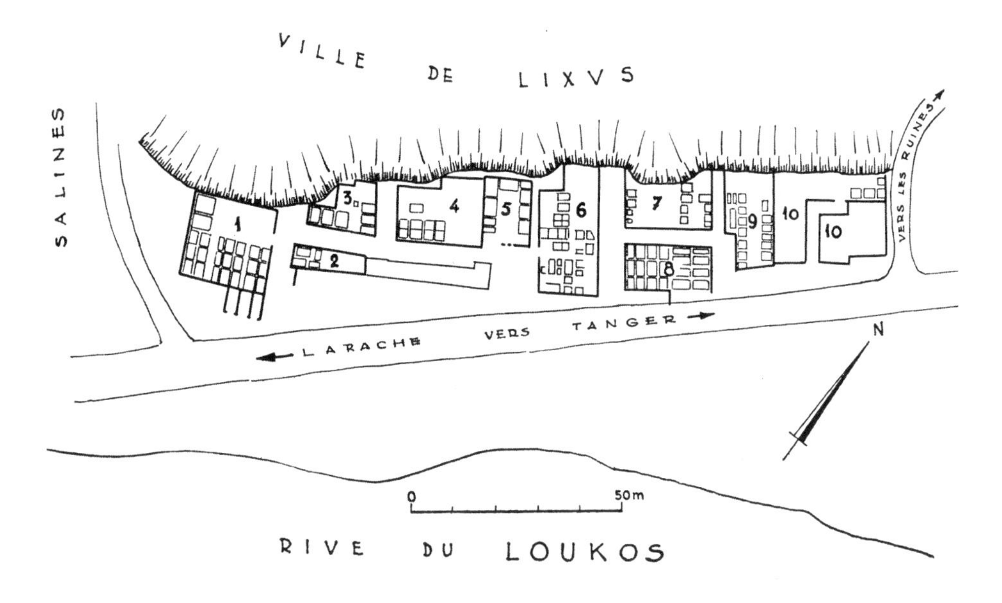

Fig. 16. The extant fish-processing complexes at Lixus (after Ponsich and Tarradell 1965, fig. 3).

study (Fig. 15). From east to west, these include Sania e Torres, Ceuta (*Septem Fratres*), Alcazarsegher, Sahara, Cotta, Tahadart, Kouass, *Lixus*, *Thamusida*, and Îles Purpuraires at Essaouira.[75]

The chronology for the use of the sites throughout *Mauretania Tingitana* is well established. Many of the installations, such as *Lixus*, Kouass, Tahadart, Cotta, Ceuta and Essaouira, began to operate in the late first century BC. As in southern Spain, the greatest period of activity in the region was in the first century AD,[76] and other sites were established at this time, including Sahara and Alcazarsegher in the Straits of Gibraltar, and possibly Sania e Torres and the *cetariae* at *Thamusida*. Mirroring the archaeological record of *Baetica* and *Tarraconensis*, the production centres of Cotta, Sahara, Alcazarsegher, *Thamusida*, and Essaouira ceased operation in the third century AD. However, Kouass and Tahadart functioned well into the fifth century or later; after hiatuses in the third century, *Lixus*' production was reduced in size and operated until the start of the fifth century and Ceuta's *cetariae* were used again in the fourth and fifth centuries. Sania e Torres' few *cetariae* could have been used continuously until the beginning of the fourth century.[77]

The processing sites of *Mauretania Tingitana* vary in size and plan, and both Cotta and Essaouira, like some sites in *Baetica*, *Tarraconensis*, and *Lusitania*, were associated with villas.[78] In the case of Essaouira, the three identified vats probably sustained nothing more than the consumption needs of the villa and its dependents. Sania e Torres, Kouass, and Sahara were never more than a few isolated *cetariae*, and probably were associated with other larger sites or towns in the region, such as Ceuta and Zilil. The sites of Cotta and Tahadart are also isolated from larger settlements or towns, but are in fact extensive complexes,[79] Tahadart being fairly reminiscent of Tróia, but on a much smaller scale.

Some fish-processing installations were located close to large settlements, and those at *Lixus* were the largest in the western Mediterranean (Fig. 16).[80] The production area at *Lixus* is located near the shore of the Oued Loukkos, just outside the city walls and below the acropolis, with no other attached buildings or residences. This situation is also similar to that at *Thamusida*, where several *cetariae* were located outside the city walls on the shore of the Oued Sebou.[81] At *Lixus*, the processing installations consist of ten extant, closely-spaced complexes; more certainly existed in antiquity, but the construction of a modern road through the site has unfortunately eliminated more archaeological vestiges. Extant are at least 142 square and rectangular vats with a combined capacity of 1,013 cubic metres.[82] *Lixus* was the only African city with fish on its coins, and these were fashioned in the style of *Gades* and *Abdera*, with fish forming columns of a temple on the reverse.[83]

3.3.1 Cotta

The most completely excavated fish-processing site in *Mauretania Tingitana* is Cotta. Located just a few kilometres south of Cap Spartel, the promontory that forms the western entrance to the Straits of Gibraltar, Cotta sits just above a wide beach on Morocco's north Atlantic coast. A small stream, Oued Khil, is located just to the north of the site, and near the installation are a small villa and temple. Cotta began operating in the first century BC and ceased functioning in the third century AD.[84]

The general plan of Cotta is very similar to those in other installations throughout the region, such as *Lixus*, *Baelo*, and Tróia (Fig. 17). The complex at Cotta is one large building, facing the sea and covering 2,240 m^2. There is a large preparation area to one side of the building and storage areas at the back and opposite side of the building. In the central room of the building are twelve large and four small *cetariae*, arranged in a U-shape around a paved preparation area. Under this area is a cistern with a volume of 86 m^3 (Fig. 18). Adjacent to this area and next to the complex entrance is a small room with a furnace and hypocausts.[85] Adjoining baths, an olive press, and attached peristyle house also compliment the complex at Cotta.[86]

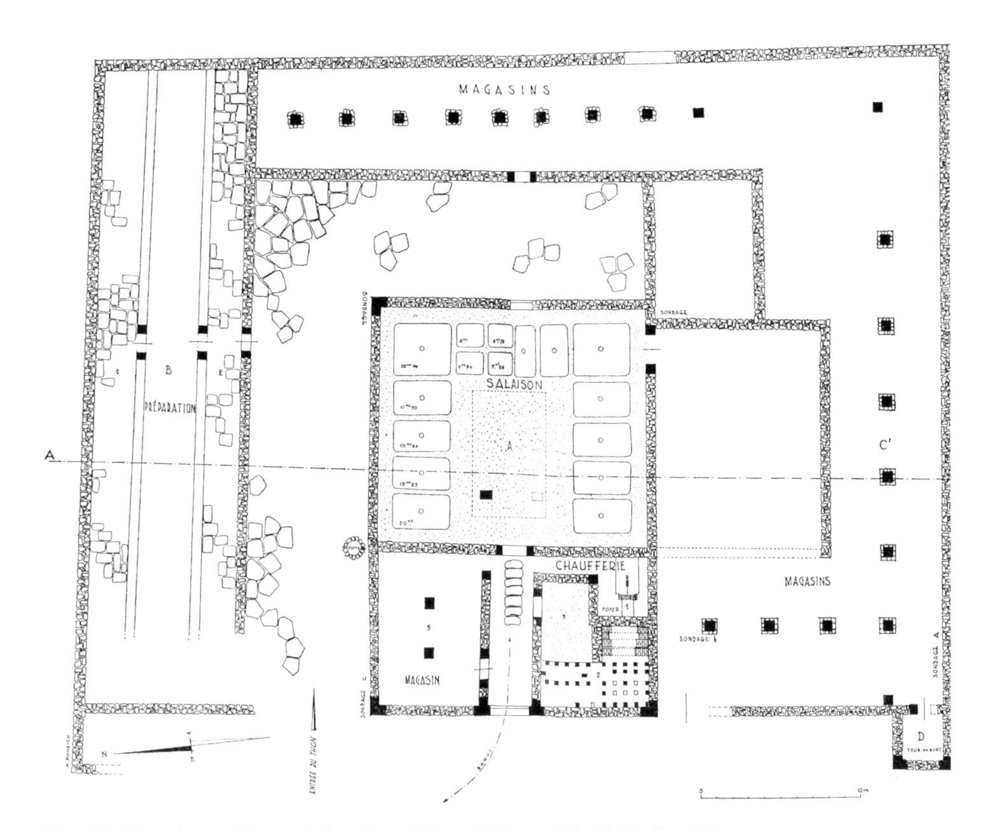

Fig. 17. The plan of Cotta (after Ponsich and Tarradell 1965, fig. 36).

On the south-western corner of the factory building, facing the sea, is a square addition, thought by M. Ponsich and M. Tarradell to have been a watchtower, or more specifically, a tunny watchtower (θυννοσκοπεíον). Such towers, mentioned by Strabon (5.2.6; 5.2.8; 17.3.16), were utilised by lookouts, who could spot the migration of tunny by observing changes in the colour or surface pattern of the ocean from their dense schools.[87]

The *cetariae* at Cotta lie flush with the floor of the building and are over 2 m deep, holding an estimated volume of 258 m^3. Some are rectangular in shape, and two of the *cetariae* are square, measuring 3.5×3.5 m. At the bottom, these vats have small circular pits or cuvettes to assist in cleaning between batches. As the sun unwontedly accelerated the evaporation process in making fish sauces, the facility at Cotta, as also documented at *Baelo*, had a roof covering it, supported by pillars. However, there were most likely windows or openings in the walls to allow for ventilation. The small furnace near the entrance of the complex fed the hypocaust system for the artificial heating of fish-sauce mixtures, and the small ceramic pots with handles and spouts used for this process, *marmites*, were found in abundance at the site.[88]

Fig. 18. The cetariae *of Cotta around the central workspace. The workspace floor (upper left) has now given away, revealing the cistern (photo: A. Trakadas).*

Although Cotta represents a smaller production output than some of the individual complexes at *Lixus*, *Baelo*, and Tróia, it is an example of a purpose-built and self-sufficient fish-processing factory. The complex at Cotta is one large unit, and the central production building of the site was laid out for the efficient processing of salted fish products. An olive press was also installed in the building, probably producing olive oil for consumption by the workers of the site. The small temple nearby, the attached baths, as well as the presence of a necropolis to the south, would suggest that the workers of the complex lived nearby and were dependent on the installation for part of, or perhaps the entire year. Cotta was most likely the industrial annex of the nearby villa, and the attached peristyle house was probably the residence of the factory's manager.[89]

4. *Features of fish-processing sites in the western Mediterranean*

Throughout the southern Iberian Peninsula and north-western Africa, the remains of the fish-processing sites used during the Roman period reveal a surprising amount of homogeneity. This is demonstrated not only by the specific topographical situation of each installation, but also in the constructional details of the *cetariae* themselves and other necessary features of the complexes, such as heating facilities, water supplies, and kilns.

4.1 *Topography*

4.1.1 *Marine resources*

There is abundant sea life in the western Mediterranean and eastern Atlantic, as well as in the Straits of Gibraltar, which connects the two and serves as the major migratory route for many marine species.[90] The breeding cycles of different fish, and their migratory routes, which tended to follow the prevailing currents, were understood in antiquity,[91] and ancient writers often named specific regions in the western Mediterranean that were plentiful in fish. Capturing mackerel in the region during migratory passages is mentioned by Pliny (*HN* 9.49); Strabon (3.2.7) mentions that murry and the largest surmullets came from Spain, and that in Portugal, the Tagus River was rich in fish (3.3.1). The entire Turdetanian seaboard was also praised by Strabon (3.2.7) as being particularly rich in marine life.

Fish-processing sites throughout the western Mediterranean were uniformly sited along the coasts or major rivers of southern Spain, Portugal, and northern Morocco, but the zones where the installations were located also reflect the proximity of rich fishing grounds.[92] Locating processing sites near these grounds would considerably shorten the time between catching and processing, limiting the extent of decomposition of the catch. Fishermen could, in many instances, deliver their catches directly to processing sites, and, as Manilios (*Astronomicon* 5.656-681) describes, with the particular location of these sites, fishermen could come to shore near the installations and start to clean their catches of tunny, cutting it into pieces and wasting no portion.

4.1.2 *Water supply*

Part of the essential requirements for the processing of fish was fresh water, which would serve for washing fish, preparing brine, and cleaning the processing installations themselves. Almost all of the processing sites in the western Mediterranean are located near naturally-occurring bodies of fresh water, such as rivers or streams, but many sites also developed systems for making sure a necessary amount of fresh water was always on hand. This includes wells, which are present at many sites, but also cisterns and aqueducts.[93] Sites on islands, such as Îles Purpuraires at Essaouira, *Scombraria*, and Ilha do Pessegueiro, had cisterns and wells, but so did many other sites on the mainland.[94] The site of *Lixus* had two buried cisterns,[95] and a cistern is also associated with Cotta.[96] At Guincho, on the Atlantic coast west of Lisbon, there is a large elevated tank with a connecting reservoir and channels.[97] Quinta da Comenda had a canal for water,[98] as did Ceuta, which was connected to a nearby cistern.[99] Kouass also had wells and an aqueduct over 750 meters long with a subterranean portion that terminated in a collecting pool.[100]

4.1.3. *Salt resources*

In the Roman world, fish could be processed in two basic ways: the flesh could be cut up and salted, forming *salsamenta*, or the leftovers and/or small fry could be macerated with salt and fermented, forming the various liquid fish sauces (*garum, liquamen, muria, allec,* and *lymphatum*). Processing with salt was an innovative method for preserving a necessary food item in a world without any means of refrigeration, and made possible the trans-shipment of preserved fish and fish sauces to distant locations.[101] The processes involved with salting are described in several texts: Pliny (*HN* 31.93-95) only states that fish parts were combined with salt to make *garum*, but the ratio of fish to salt when making *garum* is described in *Geoponika* (20.46.3) as being 8:1. The method for making *salsamenta* described by Columella (12.55.4) requires square pieces of fish to be covered with salt.

As a constant supply of salt was therefore necessary for manufacturing *salsamenta* and fish sauces, many of these sites were also located near salt marshes or salt mines. In Portugal, the major fish-processing sites were located in areas where there are also major salt resources, the Sado Estuary and Algarve,[102] and the nearby coast of Turdetania is mentioned by Strabon (3.2.6) as a source of good-quality salt. Other major salting regions included Almería and Cadiz in Spain, and the Oued Loukkos basin at *Lixus* and at Kouass in Morocco.[103]

4.2 *Salting vats: cetariae*

The vats used for processing fish, called *cetariae* (Pliny, *HN* 9.92),[104] are remarkably similar and almost universal in their construction in the western Mediterranean. They were usually built flush with the ground or slightly protruding, although some were built on top of rocky promontories.[105] Usually vats were rectangular or square in shape, but did vary extensively in size and depth. The walls of *cetariae* were built of bricks and/or rubble construction, which were faced with a sealing mortar mixture of lime and small fragments of tiles or ceramics, forming *opus signinum,* occasionally called *cocciopesto* (Fig. 19).[106] The top corners of *cetariae* were rounded, and in some examples, the interior, bottom edges had a quarter-round or "ovolo", to prevent coagulation of the fish mixture in the corners and assist in cleaning the vats.[107]

Uniquely in the region, only four *cetariae* at *Baelo* and one small example at *Lixus* (in complex No. 4) are round.[108] Faced with *opus signinum*, these examples were clearly used for fish processing. However, at the installations at Calpe, Punta de l'Arenal, and Ceuta, large round holes are present in the ground near *cetariae*. At Punta de l'Arenal, the holes are cut into the rock (as

Fig. 19. Cetariae *construction:* opus signinum *facing over rubble construction, visible at Cotta (photo: A. Trakadas).*

were the *cetariae*), and at Ceuta, the 1.5 mø hole was lined with stonework.[109] These holes probably held buried dolia, which could have also been used for fish sauce production. Dolia are suggested as containers for making *garum* by Manilios (*Astronomicon* 5.679), and remains of dolia with fish bones inside, probably evidence of *allec*, have been excavated at a "garum shop" at Pompeii, and also at Tyritake, Myrmekion and Chersonesos in the Crimea.[110]

Some *cetariae* were built with a small, rounded catch basin or cuvette in the bottom, to help with cleaning. Usually the cuvettes are centred in the *cetariae*, but some examples are located in a corner. This feature appears in vats, for example, at Punta de l'Arenal,[111] Villaricos,[112] *Baelo*,[113] Portimões,[114] Olhão, Quinta do Lago,[115] Tahadart, Cotta, and Sania e Torres.[116] Another construction feature that aided in draining a *cetaria* was a small inclined conduit that passed through the wall of the vat and could be closed by a plug. The conduit would lead to a small catch basin or simply open on to the floor of the processing room. In one large rectangular vat at Alcazarsegher, a conduit drained into a semi-circular basin (Fig. 20); the same feature was present in two smaller basins at the nearby installation of Sahara.[117] At Tavira, a large *cetaria*, 4×2.80 m and 1.2 m deep, had a lead-lined conduit installed through the wall to drain the vat.[118] This feature is unique to these sites discussed here, but is also present at Rhodes in Spain (north of Barcelona), where three vats had such conduits leading to round catch basins.[119] At *Lixus* (complex No.

8), three *cetariae* were joined together by two arched passages through their walls, much like multi-chambered cisterns.[120]

Most *cetariae* were joined together in rows along the inner walls of a room or building, allowing for a central area to serve as a work place. This organization of space can be seen clearly at Cotta, *Baelo*, and Tróia, but also at other installations such as Roquetas, Adra, Villaricos, and Tahadart.[121] Workers could access the vats by walking on top of the walls between them. At many sites, such as Tróia, *Baelo*, and Tahadart, the paved preparation areas in front of the vats had sloping floors that drained to circular sumps that collected refuse.[122] At some sites, however, *cetariae* were not joined but stood independent of each other. This is the case at Punta de l'Arenal and at Praia de Angeiras, where at both sites the vats were cut into rock.[123]

That *cetariae* were left uncovered to assist in the fermentation process is humourously related by Pliny (*HN* 9.92), who describes that the uncovered tanks at *Carteia* (modern El Rocadillo) were relieved of their salted fish by a giant polyp. Four large windows are present in the extant walls of one of the installations at *Baelo*, and four windows are also present at Tahadart (in installation No. 1).[124] There is evidence of columns for supporting roofs at these two sites and Tróia and Cotta; it is assumed that roofs were necessary, to protect the mixtures from the elements, but that windows or open walls were a method of ensuring ventilation.[125]

4.3 Heating facilities

Furnaces and hypocausts often constitute the facilities of many fish-processing complexes. These were used to artificially heat fish-sauce mixtures, reducing the concentration, optimally by 2/3. This process is described by Ps.-Rufius Festus (*Brev.*) and in the *Geoponika* (20.46.1-6) as a quick method to produce *garum*, and mixtures were sometimes initially heated in small bowls with handles and spouts called *marmites*.[126] Unfortunately, in many cases, hypocausts utilised for this process are often identified in many early archaeological reports as "baths", such as at Tróia,[127] San Pedro de Alcántara, Punta de l'Arenal,[128] Senhora da Luz, and Portimões.[129] A furnace and hypocausts are present in the actual complex buildings at Cotta and Tahadart, and possibly at Kouass.[130] A furnace is also present at Sanlúcar de Barrameda,[131] and at Essaouira; in the latter it is associated with the nearby villa, but it could have also served for heating fish sauce.[132]

4.4 Kilns[133]

Some salted-fish producers, like those who made wine in antiquity, probably manufactured their own amphorae for the transhipment of their products.[134] Kilns that produced transport amphorae therefore formed a necessary part of the salted-fish industry, and many are located near or associated with several

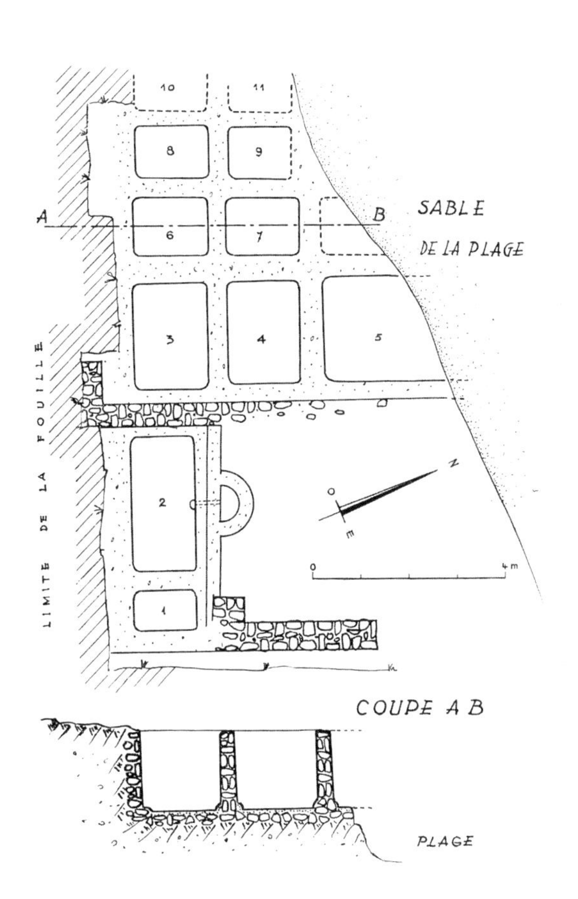

Fig. 20. The drainage conduit present in the construction of one of the cetaria *(No. 2) at Alcazarsegher (after Ponsich and Tarradell 1965, fig. 48).*

fish-processing sites throughout the western Mediterranean. At least five kilns existed in the region of Cadiz, two were near El Rocadillo,[135] and one possibly was used at Sanlúcar de Barrameda.[136] Three kilns are distributed throughout the Tagus Estuary, nine kilns in the Sado Estuary and eight in the Algarve; because of their location and products, these kilns must be associated with the fish-salting industries of *Lusitania*.[137] In Morocco, a large kiln was located at Kouass, which began operating in the fifth century BC, manufacturing Phoenician, Punic, and later Roman types, as well as imitations thereof. [138] *Lixus* also possessed kilns,[139] and a small kiln was associated with Cotta.[140]

5. *Conclusions: Chronology and organisation*

In the first century BC, Strabon (3.4.6) describes the products of Turdetania and the region around *Gades* as producing products not inferior to those from the marine-rich Pontic region. At this time, several large fish-processing sites, such as *Baelo*, Tróia, *Lixus*, and Cotta, had begun to operate. By the first century

AD, as Galen (*On the Properties of Foodstuffs* 3.30.3-6) notes, the best grade of Pontic salted fish had become second to the products of the western Mediterranean, and "Spanish" products were held in preference above all others. It is during this century that nearly all of the 88 extant fish-processing sites in the region began to operate. The western Mediterranean provinces had productive economies with markets throughout the Empire by the end of the first century, and their salted-fish products were exported to Greece, Egypt, Syro-Palestine, North Africa, Gaul, and Britain.[141] Moreover, the installations throughout *Baetica*, *Tarraconensis*, *Lusitania,* and *Mauretania Tingitana* became the major suppliers for Rome in the period from the first to third centuries AD.[142] While inexpensive processed fish may have also come to Rome from the Pontic region, it has been postulated that the fisheries there and in the eastern Mediterranean never were or ceased to be "commercially" important to Rome, although their products were certainly consumed locally.[143]

During the third century, the production of a majority of the fish-processing installations in the western Mediterranean was clearly affected. Many of the installations went out of operation; a few severely curtailed their production, or were even briefly abandoned and re-opened in a limited fashion. Installations such as *Baelo*, Tróia, and a majority of those in *Mauretania Tingitana* and *Lusitania* continued limited operations until the fifth or sixth centuries, a fact which is confirmed by Ausonius (*Letters* 25).[144] Spanish fish-sauce amphorae were still imported into Ostia in the fourth and fifth centuries, and are found on shipwrecks of the period,[145] but it is clear that there was a decline in the production and trade of the products of the western Mediterranean provinces. By the late third century in Rome, an increase in the importation of North African goods can be seen, and excavations at the "Baths of the Swimmer" in Ostia demonstrate that Africana I and II transport amphorae (thought to contain salted-fish products) start to dominate the Roman import markets.[146]

The explanation for this change in production and operations, however, cannot be conclusively tied to any one determinant. Although underlying environmental factors that affected fish catches cannot be eliminated, the impetus was certainly politically and economically charged. It has been postulated that the change was the result of an "economic crisis" and sudden in apparition, but the reason for the industry's demise in the western Mediterranean perhaps was due to the general political instability of the Empire after the death of Commodus in 192 AD, resulting in a slow economic decline over the next century.[147] Certainly, the industry was affected in *Mauretania Tingitana* by the barbarian invasions of the later third century.[148]

The numerous fish-processing installations in southern Spain, Portugal, and northern Morocco are very homogenous with regard to their topographical situations, constructional details and chronology, and imply that close ties were shared between the regions in antiquity. Not only are the fish-processing zones of the western Mediterranean connected environmentally, but during antiquity these areas were also connected culturally as Roman provinces. The geo-political cohesiveness of the western Mediterranean provinces may have caused many of the region's industries, such as wine making, olive oil production and also fish processing, to function as economic units on a certain level.[149] M. Ponsich suggests that southern Spain, especially the province of *Baetica*, dominated the region politically and economically, exerting particular control over *Mauretania Tingitana*. In this way, salted-fish products from installations in northern Morocco were probably shipped to *Gades*, under a "cooperative" or "consortium" arrangement; the products were then exported by merchants throughout the Empire under the "Gaditanian" label.[150] Such a consortium-like arrangement would, as Ponsich suggests, explain the lack of texts referring to the products of North Africa and the existence of many referring to the products of *Gades*.[151]

J.C. Edmondson also suggests that a similar scenario initially developed for the province of *Lusitania*; here, the fish-processing industry began as an adjunct to that of *Baetica*'s, and surplus products were transhipped through *Gades*, possibly even under the "Gaditanian" label. Only when Lusitanian forms of salted-fish amphorae appear outside of the region in the middle of the first century AD, is it clear that this province exercised some degree of mercantile independence. These amphorae, however, could still have been transhipped through *Baetica*, but as clearly distinguishable Lusitanian goods.[152]

Although geo-political ties certainly existed in the region in antiquity, it has not been proven, however, that commerce in salted-fish products was organised on a provincial level, or that any one consortium was able to maintain a monopoly. Imperial fish-processing sites did exist in Spain, but almost all the installations in the western Mediterranean provinces were privately owned.[153] As Robert Curtis suggests, individual operators of these various installations in the region could have functioned independently, but also could have had the opportunity to sell their products to large organisations or *societates* in southern Spain that certainly existed in *Baelo*, *Gades*, and *Carthago Nova* (the latter's *garum sociorum* subject to treatment by Pliny (*HN* 9.66; 31.93) and Martial (13.102)). From these consortia, salted-fish products could then be transhipped under one merchant or shipper's "label" for export throughout the Empire.[154]

At certain sites in the western Mediterranean provinces, fish processing did occur on a limited level, distributing goods for local consumption. However, an overwhelming majority of the installations in the region certainly demonstrate surplus production. In some instances, the installations associated with villas, especially in the Algarve region of southern Portugal, are of a scale that reflects an industrial annex of a "landed estate", serving as just one of the sources of revenue.[155] Production at such sites was most likely seasonal and took place only in the summer months.[156] The larger complex at Cotta, and those at *Baelo* and Tróia, obviously occupy much different rungs on the scale of production and were part of a more developed, year-round, urban economy. The term "industry" seems most appropriate to describe the organisation evidenced by these sites. These three sites illustrate clearly the dynamics of the fish-processing industry in the region and reflect the extent of the economic prosperity that it experienced in the first few centuries AD in the western Mediterranean.

Table 1. Key to site numbers

Spain

1. Denia (Dianium)
2. Punta de Castell
3. Punta de l'Arenal (or Jávea)
4. Calpe
5. Campello
6. Tossal de Manises (Lucentum)
7. The island of Tabarca
8. Santa Pola (Portus Illicitanus)
9. Mar Menor
10. Cartagena (Carthago Nova)
11. Scombraria
12. Mazarrón
13. Villaricos (Baria)
14. Torre García
15. Almería (Portos Magnos)
16. Ribera de la Algaida
17. Roquetas
18. Guardias Viejas
19. Adra (Abdera)
20. Almuñúncar (Sexi)
21. Torrox
22. Cerro del Mar (Maenuba)
23. Málaga (Malaca)
24. Guadalhorce
25. Torremolinos
26. Fuengirola
27. San Pedro de Alcántara (Silniana)
28. El Rocadillo (Carteia)
29. Algeciras (Iulia Traducta)
30. Villavieja (Mellaria)
31. Bolonia (Baelo)
32. Barbate (Baesippo)
33. Trafalgar
34. Puerto Real
35. Cadiz (Gades)
36. Sanlúcar de Barrameda

37. Cerro del Trigo
38. Huelva (Onuba)

Portugal

39. Quinta do Lago
40. Quinta do Muro
41. Cacela
42. Tavira (Balsa)
43. Alfanxia
44. Olhão
45. Faro (Ossonoba)
46. Loulé Velho
47. Quarteira
48. Cerro da Vila
49. Armação de Pera
50. Ferragudo
51. Portimões
52. Boca do Rio
53. Mexilhoeira Grande
54. Vau
55. Paul
56. Senhora da Luz
57. Burgau
58. Salema
59. Ilheu de Baleeira
60. Ilha do Pessegueiro (Poetanion)
61. Sines
62. Tróia
63. Alcácer do Sal (Salacia)
64. Santa Catharina
65. Senhora da Graca
66. Pedra Furada
67. Cachofarra
68. Setúbal (Caetobriga)
69. Comenda
70. Rasca
71. Creiro
72. Alfarim
73. Casilhas
74. Lisbon (Olisipo)
75. Guincho
76. Garrocheira
77. Atouguia
78. Praia de Angeiras

Morocco

79. Sania e Torres
80. Ceuta (Septem Fratres)
81. Alcazarsegher
82. Sahara
83. Cotta
84. Tahadart
85. Kouass
86. Lixus
87. Thamusida
88. Îles Purpuraires at Essaouira

Notes

1 Curtis 1991, 6-9.

2 Ponsich 1970, 28-66.

3 Tarradell 1968, 96; Sutherland 1939, 101-102; Ponsich 1968, 12.

4 Étienne 1970, 298-299.

5 Ruiz Mata 2002, 160-161; Aubet Semmler 2002, 201-203, fig. 1.

6 Fish bones and shells were excavated in eighth-century Phoenician levels at *Sexi* and Toscanos (Pellicer Catalán 2002, 57; Uerpmann 1972; Lepiksaar 1973; Molina and Huertas 1985, 26), as well as the Phoenician and Punico-Mauretanian layers (seventh-fourth centuries BC) at *Lixus*, Morocco (Grau Almero, *et al.* 2001, 204-220).

7 Ponsich and Tarradell 1965, 109-111. However, fish also appear on the coins of cities not situated along coasts: *Ituci*, *Asido*, *Ilipa* and *Caura* (Blázquez, *et al*. 1978, 234).

8 M. Aubet (1987) suggests instead that agriculture and domestication of animals in the small coastal hinterland was the initial reason for Phoenician settlement of *Sexi* and *Abdera*, and later they served as centres of navigational support for vessels from *Gades*.

9 Muñoz Vicente, *et al*. 1988, 488-490; de Frutos et al. 1988.

10 Muñoz Vicente, *et al.* 1988, 490-496.

11 Types and parallel types of these amphorae have also been found at *Sexi*, Ibiza, Kouass (northern Morocco), and Sicily, dating from the end of the sixth to the beginning of the fourth centuries BC. Pellicer Catalán 2002, 74-76; Purpura 1982.

12 Williams 1979, 111-114, 117-118.

13 *CAF* 1, Eupolis, 310, fr. 186; *CAF 2,* Nikostratos, 220, fr. 4; Antiphanes, 43, fr. 77.

14 Ponsich and Tarradell 1965, 113-114. Two main areas of fish processing during the Roman period in Portugal, the Algarve and the Sado Estuary, also correspond closely to the areas of early Phoenico-Punic influence (Garcia y Bellido 1942a, 82-93). The same is also true along the Atlantic coast of Morocco, after the initial Phoenician colonization in the eighth century near Cotta, *Lixus*, and eventually Îles Purpuraires at Essaouira (Ponsich 1970, 76-81, 106-165; Aubet 1993, 135-137, 219).

15 Curtis 1991, 48, n. 28.

16 Clément 1999, 113.
17 The amphorae are Dr. 12 and Dr. 7-11 types; Tailliez 1961.
18 Hesnard 1980, 141, 146-148.
19 See Curtis 1988, 209, n. 3, for a discussion of Plautus and Spanish products in the second century BC.
20 Ponsich 1988, 30-43; Blázquez, *et al.* 1978, 392.
21 Ponsich and Tarradell 1965, 81-89; Ponsich 1988, 169-218; Curtis 1991, 48-50, 104; García y Bellido 1942b, 2; Martin and Serres 1970; Cara Barrionuevo, *et al.* 1988; Sotomayor 1971; Loza Azuaga and Beltrán Fortes 1988, 996-997. I have purposely limited this list to include a cohesive unit of only the southern coast of Spain including the Alicante region; other Spanish sites in the Mediterranean do exist on the island of Ibiza, and further north on the Catalonian coast at Barcelona and Rhodes.
22 Bonneville, *et al.* 1984.
23 Martin and Serres 1970, 87.
24 Ponsich and Tarradell 1965, 81, 85, 89, 118-119.
25 Ponsich 1988, 198, 233-234; Curtis 1991, 59; Cara Barrionuevo, *et al.* 1988, 933.
26 Martin and Serres 1970, 39-41, 84; Martin 1970, 144-151.
27 There might also be an example near Barbate (Curtis 1991, 53, n. 42).
28 Martin and Serres 1970, 14.
29 Nicolaou and Flinder 1976, 137-139; Ponsich and Tarradell 1965, 81-82; Martin and Serres 1970, 17.
30 Corcoran 1957, 125-127; Zeepvat 1988, 17-18; Gazda and McCann 1987, 141-155.
31 Ponsich and Tarradell 1965, 83; Gorges 1979, 181; Martin and Serres 1970, 18.
32 Ponsich and Tarradell, 1965, 86-88; Ponsich 1988, 192-199.
33 See Gajdukevič 1971. By the Byzantine period, such situations had changed; at this time, fish-processing sites were required to be placed at least three stades from settlements, due to the smell (Curtis 1991, 188-189, n. 18).
34 Domergue, *et al.* 1974, 9-10.
35 Martin-Bueno, *et al.* 1984; Dardaine and Bonneville 1980, 391.
36 Ripoll López 1988, 484-485.
37 Dardaine and Bonneville 1980, 398; Ponsich 1988, 198; Pelletier 1988, 805-809.
38 Ponsich 1976; Pelletier 1988, 804.
39 Domergue, *et al.* 1974, 22; Ponsich 1988, 192-199.
40 Ponsich 1988, 78, 197-198.
41 Ponsich 1988, 40, fig. 14.
42 Curtis 1991, 52, n. 39.
43 Curtis 1991, 51.
44 Dardaine and Bonneville 1980, 386-388.
45 Bonneville, *et al.* 1984, 470-476; Ponsich 1976, 72-74.
46 Ponsich and Tarradell 1965, 86-87.
47 Knapp and Stanley 2000, 415.
48 de Figueiredo 1906, 109-113.
49 Ponsich and Tarradell, 1965, 109-111; Edmondson 1987, 100.
50 Edmondson 1987, 105-106; Clément 1999, 119.
51 Ponsich 1988, 221-228; Ponsich and Tarradell 1965, 89-90; Edmondson 1987, 255-269; Cleto 1995-1996; de Figueiredo 1906.
52 de Figueiredo 1906, 109-111.
53 Edmondson 1987, 108.

54 Marques da Costa 1960, 16; Tavares da Silva 1980; Edmondson 1987, 108-109.
55 Soares 1980; Edmondson 1987, 263.
56 de Figueiredo 1906, 113-114.
57 Cleto 1995-1996, 24-27; Edmondson 1987, 269; Tavares da Silva and Soares 1993.
58 Gorges 1979, 480-483; Gil Mantas 1999, 147, 151.
59 Gil Mantas 1999, 151; de Alarcão 1988a, 88.
60 Dias Diogo and Cavaleiro Paixão 2001, 117; Étienne, *et al.* 1994, 11-13.
61 Étienne, *et al.* 1994, 27-40; Edmondson 1987, 108, 267-268; de Alarcão 1988b, 130-131.
62 de Alarcão 1988b, 130.
63 de Almeida, *et al.* 1978, 2; Edmondson 1987, 124-125, fig. 5.5.
64 Edmondson 1987, 124.
65 Étienne, *et al.* 1994, 70-76.
66 Étienne, *et al.* 1994, 78-82.
67 Étienne, *et al.* 1994, 82-88.
68 Étienne, *et al.* 1994, 88-92.
69 Edmondson 1987, 124.
70 Edmondson 1987, 268.
71 Edmondson 1987, 131; de Alarcão 1988b, 130. For fishing implement manufacture: Ribeiro 1971, 398-399.
72 Étienne, *et al.* 1994, 69, 161; Dias Diogo and Cavaleiro Paixão 2001, 117; Marques da Costa 1930-31; de Alarcão 1988a, 43; Gil Mantas 1999, 147.
73 Grau Almero, *et al.* 2001, 204-220; Ponsich 1970, 28-66.
74 Aubet 1993, 135-137, MacKendrick 1980, 290-293; Ponsich 1970, 183-194.
75 Ponsich and Tarradell 1965, 9-77; Rebuffat 1977, 284-285; Jodin 1967, 68-71, 256-262; Bravo Perez 1980.
76 Ponsich and Tarradell 1965, 39, 68; Ponsich 1988, 103-136; Jodin 1967, 261-262; Ponsich 1967, 375; Jodin 1957, 19; Bernal Casasola and Pérez Rivera 1999, 28-29.
77 Ponsich and Tarradell 1965, 9-77, 118-119; Rebuffat 1972, 53; Bernal Casasola and Pérez Rivera 1999, 28-32, 46; Gozables Craviota 1997, 128; Jodin 1967, 261-262.
78 Ponsich and Tarradell 1965, 56-61; Jodin 1967, 48-56.
79 Ponsich and Tarradell 1965, 10, fig. 3, 73, fig. 48, 40-68.
80 The sites in Ceuta that show evidence of fish processing (Gran Via, Queipo de Llano, Parque de Artillería and El Paseo de las Palmeras) could possibly be connected, and represent one large production centre, making it the largest in *Mauretania Tingitana* and therefore the western Mediterranean. However, since excavations have been sporadic and modern development extensive, there is no way to ascertain if these four sites are in fact connected (Bernal Casasola and Pérez Rivera 1999, 40).
81 Rebuffat 1972, 53.
82 Ponsich and Tarradell 1965, 9-37.
83 Ripoll López 1988, 483-485.
84 Ponsich and Tarradell 1965, 55-57.
85 Ponsich and Tarradell 1965, 55-68.
86 Ponsich 1970, 283-284.
87 Ponsich and Tarradell 1965, 60-61. At the present time, tunny still migrate by Cotta and are caught in *madraba* nets by Moroccan fishermen just south of the site, where they are harvested in late July.

88 Ponsich and Tarradell 1965, 55-68.
89 Ponsich 1970, 206-208, 276.
90 Ponsich 1988, 30-43; Ponsich 1976, 70-71; Corcoran 1957, 64-65.
91 Ponsich and Tarradell 1965, 93-98.
92 Even though the coasts and waterways of southern Spain and Portugal have undergone some topographical alteration since the Roman period, particularly with regard to the silting of river mouths and the alteration of river courses, these sites in antiquity would have been located above the high-tide or flood zones (Knapp and Stanley 2000, 415; Hoffman and Schulz 1988, 53, 59; Mabesoone 1963).
93 Ponsich and Tarradell 1965, 102.
94 Jodin 1967, 71-75; Cleto 1995-1996, 32-36.
95 Ponsich and Tarradell 1965, 35.
96 Ponsich and Tarradell 1965, 57.
97 Edmondson 1987, 263.
98 Edmondson 1987, 265.
99 Bernal Casasola and Pérez Rivera 1999, 28-65; Gozables Craviota 1997, 128.
100 Ponsich 1967, 393-404.
101 Curtis 1991, 15-16.
102 Rau 1984, 40-46.
103 Martin and Serres 1970, 8; Ponsich and Tarradell 1965, 100-101.
104 *Cetaria* could also mean an enclosure for tunny, see Horace *Sat.* 2.5.44 and Pliny *HN* 9.49; 37.66; Corcoran 1957, 126.
105 de Figueiredo 1906, 118-119, fig. 4.
106 Edmondson 1987, 122.
107 Edmondson 1987, 122; Cara Barrionuevo, *et al.* 1988, 920.
108 Ponsich 1988, 40, fig. 14; Ponsich and Tarradell 1965, 18-19, fig. 8.
109 Curtis 1991, 54; Martin and Serres 1970, 85; Bernal Casasola and Pérez Rivera 1999, 50.
110 Curtis 1991, 92-94, fig. 6, 123, n. 55.
111 Martin and Serres 1970, 41, fig. 20.
112 Cara Barrionuevo, *et al.* 1988, 931, fig. 9.
113 Dardaine and Bonneville 1980, 382-383, fig. 6.
114 de Figueiredo 1906, 116.
115 Étienne, et al. 1994, 111, figs. 38-39.
116 Ponsich and Tarradell 1965, 44, 60-61, 76.
117 Ponsich and Tarradell 1965, 68-73, figs. 45, 48.
118 de Figueiredo 1906, 118.
119 Nolla-Brufau 1984, 437-442, 453, fig. 15.3.
120 Ponsich and Tarradell 1965, 31, fig. 18.
121 Cara Barrionuevo, *et al.* 1988, 922, 930-931.
122 Étienne et al. 1994, 76; Curtis 1991, 51; Ponsich and Tarradell 1965, 44, fig. 27.
123 Martin and Serres 1970, 39-41, fig. 3; Gil Mantas 1999, 152, fig. 4.
124 Ponsich 1976, 71-73, fig. 1; Ponsich and Tarradell 1965, 44, fig. 27.
125 Blazquez, *et al.* 1978, 234; Martin-Bueno, *et al.* 1984, 491.
126 Ponsich and Tarradell 1965, 105-6; Ponsich 1970, 287-288.
127 de Alarcão 1988b, 130-131.
128 Ponsich and Tarradell 1965, 103; Martin and Serres 1970, 85.
129 Edmondson 1987, 256-257.
130 Ponsich and Tarradell 1965, 57-60 (Cotta), 43-44, fig. 27 (Tahadart), 39 (Kouass).

131 Curtis 1991, 56.
132 Jodin 1967, 257.
133 Broadly, the types of transport amphorae associated with western Mediterranean salted fish and fish sauces are: Dr. 7-11, Beltrán I, Dr. 12-13, Dr. 14, Dr. 38-39, Beltrán IIB, Almagro 50, Almagro 51a-b, and Almagro 51c (see Peacock 1974; Parker 1977).
134 Peacock and Williams 1991, 73-76.
135 Peacock 1974, 236-242.
136 Esteve Guerrero 1952, 127.
137 Edmondson 1987, 158-159, fig. 6.2; de Alarcão 1988a, 86-87, fig. 40; Parker 1977.
138 Ponsich 1967, 376-385; Kbiri Alaoui 2000.
139 Ponsich 1981, 24.
140 Personal communication, Dr. A. Elboujaday, Délégacion de la cultura de Tanger, June 2002.
141 Curtis 1991, 64; Curtis 1988b; Cotton, *et al.* 1996.
142 Curtis 1991, 180.
143 Corcoran 1957, 85.
144 Ponsich 1988, 198; Edmondson 1987, 108-109, 189; Tarradell 1955; Ponsich and Tarradell 1965, 115-117; Curtis 1991, 60.
145 For example, the Sud-Lavezzi 1 wreck with cargo of Almagro 51 amphorae dating to the fourth or early fifth century AD (Liou 1982).
146 Panella 1972, 101-104.
147 Reece 1981; Tarradell 1955; Ponsich and Tarradell 1965, 115-117; Curtis 1991, 60.
148 Keay 1984, 553; Whittaker 1994, 151.
149 Ponsich and Tarradell 1965, 4; Ponsich 1988, 30-43; Ponsich concedes, however, that this was not the case with murex dye (Ponsich 1970, 336).
150 Ponsich 1970, 336; Ponsich and Tarradell 1965, 99.
151 Ponsich 1975, 680.
152 Edmondson 1987, 189-190.
153 Curtis 1991, 181.
154 Étienne 1970; Curtis 1991, 62-63; Curtis here also discusses in detail the commerce of fish products as demonstrated by epigraphic evidence and *tituli picti*. For tertiary levels of distribution, see Curtis, 1988.
155 Edmondson 1987, 135-136.
156 Curtis 1991, 56.

The Technology and Productivity of Ancient Sea Fishing

Tønnes Bekker-Nielsen

1. *The nature of the sources*

The volume of ancient literature and inscriptions relating to saltwater fishing is not large. This reflects the social context of sea fishing: it was no profession for gentlemen – it does not figure in the writings of the elite; it was not tightly controlled by the state – there are few references to fishing in legal texts;[1] its practitioners were not wealthy – there are few gravestones or epitaphs naming fishermen. The scarcity of our evidence does not reflect a prejudice against fish and fishy matters in general – as shown by Curtis and Wilkins in preceding chapters of this volume, there is a considerable volume of texts relating to processing and consumption of fish, as well as a fair amount of evidence for freshwater fishing, fish-traps, fish-ponds etc. If we were to judge the relative importance of ancient saltwater and freshwater fisheries on the basis of literary sources alone, we might be misled to conclude that freshwater fish played a far greater role in the economy and the diet than sea fish.[2]

We do, however, have one major treatise on sea fishing, the *Halieutika* of Oppian, composed between 177 and 180.[3] The *Halieutika* is a Greek poem of more than 3500 hexameters and preserved in its entirety; for good measure, the last half of a late Roman prose paraphrase has also been handed down to us.[4] In some respects the *Halieutika* can be compared with the agricultural manuals of Varro and Columella, but the differences between these and the work of Oppian are more telling. Whereas the farm manuals are written by or dedicated to owners of agricultural land, it is quite clear that Oppian was not himself a sea fisherman,[5] and the poem is dedicated to the emperor Marcus Aurelius – supporting the notion that by this time, if not before, sea fishing rights were in principle imperial property enjoyed by his subjects at the emperor's discretion.[6] Another striking difference is that whereas the agricultural writers often comment on the economic aspects of farming: choosing the most efficient crops, getting the produce to market etc., Oppian never discusses economic matters such as the price of fish or tools, the relative efficiency of different fishing methods or how the catch is shared among the fishermen after the day's work.

The form and literary style of the *Halieutika* raise a number of disturbing questions about the nature of the information it provides. First, the use of the hexameter means that technical terms or names of certain species of fish may have been excluded because they did not scan. Second, like many other Greek writers of the second-century AD, Oppian is strongly influenced by the Hellenic revival known as the second sophistic, a retrospective literary movement striving to re-establish or reinvent Greek culture as it had been in the distant, glorious past. He draws on literary models of the classical period and may also have derived some of his factual information from writers of the fifth and fourth centuries BC, but since Oppian gives no source references, we cannot tell. Much of his information may be taken from a lost work by Leonidas of Byzantium (*fl.* c. 100 AD), other parts are clearly dependent on Aristotle. In short, it would be dangerous to assume that Oppian describes the fishing practices of his own day; his information may well be outdated by many centuries. There may also have been important regional variations in fishing technique. Oppian himself hailed from Cilicia[7] but there is nothing to suggest that he describes the fishing practice of his native region and he gives only a few examples of local fishing techniques – such as the Thracians' use of a beam with multiple tridents to catch tunny (see below p. 89).

The amount of direct archaeological evidence for sea fishing is not large, either. In exceptional environments such as Herculaneum or Egypt, remains of fishing nets have been found; elsewhere only implements made of inorganic materials, such as hooks and net sinkers, have survived. Some wrecked fishing boats may still be lying on the seabed, awaiting discovery or publication, but it is on *terra firma* that we find the most abundant evidence for fishing: the tanks used for processing saltwater fish, the containers used for shipping the finished product (cf. the contributions by Trakadas, Højte, Lund and Gabrielsen in this volume) and pictorial representations of sea fishing and fishing boats. In the Classical and Hellenistic period, such depictions are rare, but with the advent of polychrome mosaic in the first and second century AD, fishing scenes become popular, especially in Sicily and North Africa (cf. Bekker-Nielsen 2002b).

Given that our sources are diverse and widely scattered, the outcome of any attempt to describe ancient sea fishing and its productivity will to some degree be determined by our preconceived ideas about the nature of ancient society and its economy; and in recent decades, the dominating paradigm has been that of the Cambridge or "primitivist" school inspired by the work of the late Sir Moses Finley. Thomas W. Gallant's slim monograph *A Fisherman's Tale* (1985) is a work in this tradition and one of the few recent studies of ancient fishing. Gallant concludes that fishing played a minor role not only in the economy of ancient society as a whole, but even within the fisher's own *oikos*. These conclusions are based partly on modern fishing statistics, partly on technological arguments. Gallant claims that ancient sea fishing was incapable of supplying large amounts of fish because the implements

Fig. 1. Fishing with a casting-net from shore, Oman, January 2002 (photo: Jørgen Christian Meyer).

were primitive and the most efficient of the tools available, the fishing-net, was never used from boats: net fishing was "a completely shore based technology".[8] Drawing on data from nineteenth- and twentieth-century fishing statistics, he further argues that the fishing technology of the ancients would have produced risible catches, no more than a few kilos per day; so little that, under normal circumstances, it would barely support a fisherman's family or make sea fishing a viable full-time occupation. In Gallant's view, fishing formed a complement to farming, a supplementary source of nutrition and income when the harvest failed. The relevance of modern fishing statistics to antiquity is discussed by Jacobsen elsewhere in this volume; the present paper aims to examine the question of fishing productivity in the light of fishing technology.

2. *Fishing from shore and fishing from boats*

Gallant's argument for the inefficiency of ancient fishing technology rests, *inter alia*, on the assumption that net fishing did not take place from boats, but only from shore. Shore-based net fishing can, however, be highly efficient. During a field trip to Oman in 2002, Jørgen Christian Meyer observed fishermen working from the shore of the Persian Gulf with casting nets (Fig. 1). Although modern nets are made from nylon or other man-made fibres instead of flax,

Fig. 2. The result of one throw of the casting-net from shore (photo: Jørgen Christian Meyer).

the technology is similar to that used in antiquity; and in a single throw of the net, the fisher may take fifteen to twenty kilos of fish (Fig. 2).[9] Oppian's description of fishing for tunny along the coast, quoted below (p. 92-93) also suggests that fishing from shore could be quite productive.

While fishing boats[10] as well as nets[11] are mentioned in Greek literature and were no doubt used together from an early date, our present evidence for the combination of nets and boats dates from the first century AD onwards.[12]

The extant sources have very little to say about the size and range of ancient fishing boats. From a passage in Xenophon's *Hellenika* (5.1.23) mentioning fishermen returning to the Piraeus at dawn, we know that by the fourth century BC, if not before, fishing boats were large enough to range into the Saronic gulf and the fishermen sufficiently confident to navigate at night.

Two literary passages describe rulers travelling in fishing-boats: Xerxes crossing the Hellespont on his retreat from Greece in 479, in Justin's *Epitome* of Trogus, and Caesar attempting, unsuccessfully, to cross the Adriatic in 48 BC, as described by Lucan and Ammianus. On closer inspection, they have little historical value, but provide some useful insights into the general perception of fishing vessels in the Roman world.

While the near-contemporary Herodotos (8.117) merely tells us that the Persian army crossed "in ships" (*nêusi*), "as the bridges had been destroyed in a storm", Trogus elaborates on the dramatic change in the fortunes of Xerxes, illustrated by his choice of transport: the Great King is forced to use a fishing boat to traverse the Hellespont where, on his outward journey, he rode on a bridge of ships:

> Ubi cum solutum pontem hibernis tempestatibus offendisset, *piscatoria scapha* trepidus traiecit. Erat res spectaculo digna et aestimatione sortis humanae, rerum uarietate miranda in exiguo latentem uidere nauigio, quem paulo ante uix aequor omne capiebat
>
> Having found the bridge broken down by the winter storms, he crossed in the utmost trepidation in a fishing-boat. It was a sight worth contemplation for judging of the condition of man, so wonderful for its vicissitudes, to see him shrinking down in a little [fishing] boat, whom shortly before the whole ocean could scarcely contain (Justin, *Epitome*, 2.13.9-10, ed. M.-P. Arnaud-Lindet, trans. J.S. Watson)

In a similar manner, the story of Julius Caesar's abortive attempt to cross the Adriatic in 48 BC was transformed and dramatised. According to Plutarch (*Caesar*, 38) he chose a boat "of twelve oars" while in Appian's *Civil War* (2.56) Caesar sends his servants to fetch "a fast boat".

The poet Lucan tells a different and more dramatic story of Caesar walking alone on the beach in the darkness and finding a small boat whose owner is sleeping in his simple cottage nearby (*Pharsalia*, 5.504-524).[13] Ammianus, who had read Lucan,[14] reproduces this version and specifically identifies the boat-owner as a fisherman: *alium anhelante rabido flatu uentorum lenunculo se comisisse piscantis* (16.10.3, ed. Seyfarth): "another [i.e. another emperor than Constantius II, whom Ammianus is comparing unfavourably to his predecessors] in the middle of a raging gale entrusted himself to the small boat [*lenunculus*, dim. of *lembus*] of a fisherman".

On one point, then, these texts are unequivocal: to the average Roman (and presumably also the average Greek) a fishing boat was a *small* boat and not for the faint-hearted. In the opening verses of the *Halieutika*, Oppian describes "fearless" (*aphrastoi*) fishers taking to the sea in "tiny" (*baioi*) wooden fishing boats (*Hal.* 1.9; 1.41). Their small craft are directly contrasted with the large and comfortable boats used for "regal" fishing in waters where the fish are fed regularly while waiting for their owners to catch them.[15] The same impression is gained from a study of the pictorial evidence. Though the boats shown on the late Roman mosaics that form the main body of our pictorial evidence are stylized and their crews reduced to a few persons, they nonetheless give some impression of the relative size of fishing vessels compared to other boats.

In the mosaics, most fishing boats have no sails or masts. In the Althiburus mosaic, in effect a catalogue of ship types in pictorial form giving the name of each type (cf. fig. 4), the two types that can be identified as fishing vessels are rowboats, though other boats in the mosaic have masts (some also have stays, indicating a large sail).[16] In the "Mosaïque de la Toilette de Venus" found at Djemila (Culcul), two ships, one a warship, are shown with square sails but the two fishing-boats working a seine in the opposite border of the mosaic are rowboats.[17] A fishing scene forming part of a third-century African mosaic showing Bacchus fighting pirates[18] is unusual in showing three fishermen working from a large, square-rigged boat, while a fourth-century mosaic from Carthage shows two persons fishing from a boat with a mast and two stays.[19] Some boats may have had a small mast and a sail that could be raised if the wind was favourable for going to and from the fishing grounds.[20]

3. *Getting the catch ashore*

Ancient fishermen could only range over a limited area, restricted not only by the limited size of their craft but also by the short time within which the catch must be brought to market. This critical time frame could be expanded by gutting the fish immediately after the catch, by keeping them alive in baskets or creels, and by keeping the catch cool, e.g. by concentrating fishing activities in the coolest hours of the day. A passage of Galen, already quoted by Wilkins in his contribution to this volume, refers to "pickled fish or … fish that can be kept in snow until the next day".[21] Given the difficulty of obtaining snow or ice for packing, the second option was not open to our average fisherman. One notes, however, the implication that unless preserved in one way or another, fish will not keep overnight.

Fish in fact begin to deteriorate within a few hours of being caught, but if gutted immediately after the catch the rate of deterioration is reduced.[22] For table fish and some varieties of salt fish, one would assume that ancient fishermen gutted their catch on board, but the process is not described by Oppian or in pictorial sources, nor are the flocks of sea-birds that follow a boat to catch the guts as they are thrown overboard. In the production of *garum* the whole fish was used and there was no need to gut the fish at sea.

The rate of spoilage increases with the ambient temperature, so fishing at night or just before dawn, when the temperature is lowest, will increase the fisher's chances of getting his catch ashore in good condition. From Xenophon's mention of fishers coming into port in the morning, we know that as early as the fourth century BC, fishermen supplying the markets of Athens worked at night. Oppian (3.50-52) also mentions fishing late at night or early in the morning. With passive implements (ground-nets, creels, traps) set overnight, fish remain alive in the water until the fisher comes to check his nets in the morning – cf. Oppian's description (3.86-87) of nets bringing large rewards to their sleeping master. Fish could be kept alive *en route* to market in creels or well boxes (cf. p. 137 below), but although Oppian mentions creels, *kyrtoi*, several times and even explains how to make one (3.341-343) he does not mention their use for storage purposes. Roman mosaics[23] show fishers emptying creels, but the contexts imply that these, too, have been used for catching fish, not for keeping them: the creels are being emptied while the boat is still at sea.

4. *Spears and hooks*

In several passages, Oppian mentions the use of tridents to catch small sharks, swordfish, whales and young tunny (*Hal.* 3.552-554; 4.252-253). Dolphins, too, could be caught in this way. Killing a dolphin was anathema to a true Greek, but the fishermen of the Black Sea region were less sensitive in this respect. Oppian relates that "Thracians and the inhabitants of Byzantion" (*Hal.* 5.521-522) catch dolphins, and dolphin bones have been found at processing sites in Chersonesos.[24] Spears and tridents could of course also be used in shallow water and in estuaries, e.g. for catching sturgeon. Oppian mentions an ingenious device used by Thracian fishermen in the Black Sea to catch young tunny: a beam with multiple tridents attached dropped from above into the shallow water, its teeth impaling or trapping the fish (4.535-548).

Fishing with hook and line from a boat is a quite efficient method, especially if the fishermen are after large table fish. It is obviously less efficient for catching the smaller species such as mackerel, anchovies or sardines since the effort of baiting the hook remains the same regardless of the size of the fish to be caught. Efficiency also varies with the number of hooks: a line or rod with one hook is generally less productive than a line with multiple hooks.

To judge from the assertion that "line fishing is a technique incapable of output beyond a very low level";[25] T.W. Gallant apparently assumes that when fishing with hook and line, only one hook was used at a time; however, the use of multiple-hook lines is attested to not only by Oppian but by the finds of large stocks of fish-hooks on archaeological sites. In the south-eastern quarter of Chersonenos, for instance, excavators found c. 140 hooks along with 50 sinkers.[26] Such large numbers can only be explained by the use of multiple-hook lines. Assuming that the sinkers were combined with floats (of organic material such as wood or cork,[27] which will have perished) a multiple hook line could be trailed after a boat or even left overnight and drawn in next morning.

In book three of the *Halieutika* (3.78; 3.468ff) Oppian discusses the use of hooks and lines, from a boat and from shore. He gives a graphical description of how a line with multiple hooks is used to fish a shoal of saddled sea bream:

3a-b. Using the casting net from a boat, Oman, 1966 (photo: Daniel J. Bosch).

> In his hand [the fisherman] holds ready a thin rod and a thin line of light hair all untwined, whereon are strung numerous light hooks. On these he puts the same bait as before he cast in the water, and lets it down into the deep turmoil of the waves. Seeing it the Melanurus immediately rush upon it and snatch – their own destruction. (Mair's translation)

Taking Oppian's description at face value, this should be a quite efficient way of catching fish. One fully grown saddled sea bream (*Oblada melanura*) weighs 0.75-1kg.

5. *Nets and creels*

In the context of fish processing, however, our main interest must focus on nets. Nets can be adapted to almost any size of fish and a net is by far the most efficient implement for catching the smaller species often used for the production of *garum*. The widespread use of the net in antiquity is attested, *inter alia*, by the extent of the vocabulary used to describe the different types. Oppian claims that the different nets are *myria*, innumerable, but gives a short list of some of the most important types (3.79-84)

Fig. 3b.

ἀμφίβληστρον: Wurfnetz, casting-net[28]
γρίφος: Ziehnetz, draw-net
γάγγαμον: Schleppnetz, drag-net
ὑποχὴ περιηγὴς: runder Sacknetz, round bag-net
σαγήνη: Ziehgarn, seine
κάλυμμα: Decknetz, cover-net
πέζα: Grundnetz, ground-net
σφαιρῶν: Ballnetz, ball-net
σκολιὸς πάναγρος: gekrümmter Allfangnetz, crooked trawl

Some of these types can also be identified from the pictorial evidence.[29] The *amphiblêstron* is a simple and efficient device, still in use today: a casting-net that can be used either from shore (Fig. 1) or from a boat (Fig. 3a). To ensure that it sinks quickly before the fish can escape, its edges are weighted. In still waters, the edges should strike the surface of the water at the same moment (Fig. 3b); this takes some skill in casting and requires that the weights are of equal weight and evenly distributed along the edge of the net. Used from shore the casting net requires only one person, but when it is used from a boat, the mosaics typically show that at least two persons are required: one rowing, one casting the net.[30]

Fig. 4. Two fishers hauling a net (sagênê?) on board a boat (cydarum). (Drawn from the Althiburus mosaic, reproduced from Duval 1949).

The *gangamon* and the *hypochê periêgês* are likewise small nets that can be handled by one person. The *sagênê*, whence the modern word *seine*, is a larger net requiring the effort of several persons (Fig. 4), or the crews of two separate boats. In Mair's translation, Oppian describes the use of the seine as follows (4.491-496):

> Now when the fishermen behold them huddle together, they gladly enclose them with their hollow seine-nets and without trouble bring ashore abundant booty and fill with the fry all their vessels and their boats and on the deep beaches pile up heaps, an infinite abundance of spoil.

Judging from its name, the *kalymma*, "veil" would appear to be a net of unusually fine material, perhaps for catching very small fish. The *peza* or "ground" net was presumably a stationary net, weighted to the sea-bed and kept upright by floats. The *skolios panagros*, "hollow all-catching net" is rendered in Mair's translation as a "crooked trawl" but was no trawl in the modern sense of that word. The trawl is an active fishing implement dragged after the boat; it trails deeply in the water, along the sea bed. The *skolios panagros* on the other hand hangs just below the surface, suspended from cork floats along its edge.[31] Instead of moving the net itself, the fishers could attempt to shoo a shoal into the net, as Oppian describes in the case of young tunny (4.566-582, trans. A.W. Mair):[32]

> The fishers set up very light nets of buoyant flax and wheel in a circle round about while they violently strike the surface of the sea with their oars and make a din with sweeping blow of poles. At the flashing of the swift oars and the noise the fishes bound in terror and rush into the bosom of the net which stands at rest, thinking it to be a shelter; foolish fishes which, frightened by a

> noise, enter the gates of doom. Then the fishers on either side hasten with the ropes to draw the net ashore. And when they see the moving rope, the fish, in vain terror, huddle and cower together and are coiled in a mass. Then would the fisher offer many prayers to the gods of hunting that nothing may leap out of the net nor anything make a move and show the way; for if the pelamyds [tunny[33]] see such a thing, speedily they all bound over the light net into the deep and leave the fishing fruitless.

Among ancient fishing techniques, one of the most productive was to catch migrating tunny in semi-permanent nets or traps. Oppian describes the "tunny-watcher" (*thynnoskopos*) on a high hill, keeping a lookout for the approaching shoals, and the fish streaming into the stationary nets "like soldiers by the phalanx" (*Hal.* 6.637-648). In the Black Sea region, Kyzikos and Byzantion were, among others, renowned for tunny fishing. In the Mediterranean, Oppian singles out three waters as especially notable for their tunny fisheries: the Iberian Sea, the Golfe de Lion and the Sicilian Channel (3.623-627). Significantly, these three regions were also known for their *garum* production.

6. *Conclusions*

Gallant assumed that ancient net fishing did not take place from boats, and that other techniques (such as fishing with hook and line, or net fishing from shore) were inefficient. As can be seen even from this short survey of the available evidence, neither of these assumptions is tenable. Simple technologies such as lines with multiple hooks or casting-nets used from shore are capable of producing substantial catches, and nets were clearly used from boats, at least from the early Empire onwards.

In fact, the most important technological constraint on the development of Graeco-Roman fisheries was not the inefficiency of the fishing gear, but the inability to preserve fish after the catch. This limited the range of the fishing boats, since going far out of port entailed a correspondingly long return journey, by which time the catch would be spoiled. It also limited the size of boats and crews, since a larger boat would take longer to fill before it could commence the return trip.

Oppian describes seines and boats overflowing with fish as one might find on a good fishing day, but there would be other days when the catch was poor or fishing altogether impossible due to the weather. Even with advanced modern-day technology, catches remain variable and unpredictable. Catching table fish was probably never very remunerative for a fisher based in a small city (a category that includes most of the settlements along the Black Sea coast). On the few days of the year when the fisher had a windfall catch, there would be too few buyers in the local market – and the fish could not be held over until the next day. They could, however, be preserved (by drying, pickling or salting) or processed into *garum*. Compared with other strategies for obtaining food, ancient fishing technology was neither inefficient nor unproductive, and it may in fact have been overproductive in relation to potential consumption – leading in turn to the development of technologies for preservation and long-term storage of fish and fish products.

Notes

1 Ørsted 1998, 14-17.
2 Cf. that in Rostovtzeff 1957, nearly all references to "fishing" and "fish" are to freshwater fishing.
3 For the date, see *Hal.* 1.3; 2.683.
4 Published by Isabella Gualandri (1967).
5 Indeed, one may well doubt that Oppian ever went to sea in a small fishing boat. While the *Halieutika* teems with literary descriptions of the roaring waves, the rushing currents etc. there is no attempt to describe the motion of the boat itself. At the beginning of book 3, Oppian dwells on the physical stamina required of a fisherman: strength, endurance, etc. but there is no mention of seasickness; *Hal.* 3.29-40.
6 Cf. the discussion in Ørsted 2000, 19-20. The tradition that the emperor commissioned the poem from Oppian is late and dubious.
7 Cf. *Hal.* 3.7-8.
8 Gallant 1985, 25.
9 Personal communication, 23 February 2002.
10 E.g., Xenophon, *Hell.* 5.1.23
11 E.g., Hesiod, *Shield*, 213-214.
12 Bekker-Nielsen 2002b, 218-223.
13 The story of how Caesar finds the boat on the beach and wakes its master in his simple house of reeds and seaweed clearly imply that the boat-owner is a fisherman.
14 Barnes 1998, 193.
15 *Hal.* 1.57-63. Perhaps it was on such a pleasure fishing expedition that Oppian saw fishers enticing sea bream towards their lines by throwing food in the water, as described in *Hal.* 3.462-474. Cf. also 3.221-260.
16 Duval 1949; Bekker-Nielsen 2002b, No. 22; 30.
17 Blanchard-Lemée 1975, pl. 1, Bekker-Nielsen 2002b, No. 29.
18 Poinssot 1965, 224; Bekker-Nielsen 2002b, No. 18.
19 Bekker-Nielsen 2002b, No. 9.
20 In *Hal.* 3.66, Oppian explicitly advises the fisherman to set his sail when he has a favourable wind. Greek triremes similarly had a sail and mast that could be taken down when not in use.
21 Galen, *On the properties of foodstuffs 3.34* = 6.713 Kühn; see also Wilkins, above p. 24-25.
22 This and other information on fish preservation is taken from Hans Otto Sørensen's paper (not included in this volume) on the biochemistry of fish processing.
23 Bekker-Nielsen 2002b, Nos. 1-4.
24 Kadeev 1970, 16. See also Bode 2002, 13-15.
25 Gallant 1985, 25.
26 Kadeev 1970, 8.

27 For the use of cork floats, cf. *Hal.* 3.103.

28 The translations are those of A.W. Mair (English) and F. Fajen (German).

29 For a more detailed discussion of the different net types and their use, see Bekker-Nielsen 2002b; for a general survey of fishing in antiquity, see Donati and Pasini (ed.) 1997.

30 E.g., Bekker-Nielsen 2002b, No. 7; 11; 13.

31 *Hal.* 3.103, *peismata phellôn*, lit. "cork ropes", presumably cylindrical cork floats strung on a cord. Among the members of a fishing-guild at Parion on the Hellespont, one is identified as *phel[lo]chalastôn*, "loosening (?) the cork floats" (*IK* 25.6).

32 In this passage, the net is only described in general terms as *linos* ("flaxen") and *diktyon* ("webbed") but judging from the description of its use, it corresponds to the *sagênê* or the *skolios panagros*.

33 Greek *pêlamys* signifies "young tunny". The modern word "pelamyd", which Mair uses here, denotes various species of mackerel, so called from their resemblance to small tunny.

The Reliability of Fishing Statistics as a Source for Catches and Fish Stocks in Antiquity

Anne Lif Lund Jacobsen

In 1985 T.W. Gallant published an influential essay on the potential productivity of fishing in the ancient world. He concluded that: *"the role of fishing in the diet and the economy would have been, on the whole, subordinate and supplementary..."*[1] His methodological approach was original in using modern fishery data to estimate the productivity of ancient fisheries. Unfortunately his work suffered from several severe misunderstandings about ecosystems, the nature of a fishery and its biological interaction with its environment.

The purpose of this paper is to discuss the statistical background for Gallant's conclusions about fishery and the usefulness of modern catch data for historical fishery research. In order to do so, the author adopts the viewpoint of marine-environmental history, with some reference to other authors' work on ancient fisheries.

1. *Applied fishery statistics and biological literature*

Gallant's ambition is to provide some estimates about the social and economical role of fishing in antiquity. Finding the available historical sources insufficient, he chooses to base his analyses primarily on modern fishery statistics, since they have the richness and continuity that older sources lack. The different types of fishery statistic he uses to back up his argumentation can roughly be divided into two groups: Catch data and catch-per-unit-effort (CPUE) data.[2]

The first type of data consists of information on 19th century Adriatic fishery and fishery statistics, mainly from the Mediterranean and from the period 1922-69; most of the data is from the 1950s and 1960s.[3] Gallant does not tell us much about these data: whether they are total landings by country, whether they are based on commercial catch records or estimates, and how much effort was involved in the fishery. This of course makes it difficult to evaluate the quality and accuracy of the data. It is surprising that Gallant chose such a weak and incoherent statistical material since better data were available. FAO, for example, has published yearly catch data (on a national level and per species) from 1950 onwards.[4]

The content of the CPUE data is even more blurry, with some data deriving from a study of a Malaysian fishery, where there are no references to the exact catch area, species caught, or the fishing effort evolved. Gallant also seems to use some CPUE data of unknown origin. This is clear in his Figure 1 where he tries to estimate the extent to which the use of different types of fishing gear could provide sufficient fish for the daily diet of the fisherman himself.[5] Such a calculation has to be based on some sort of CPUE, but no references are given. It goes without saying that estimates such as those given in Gallant's Figure 1 are highly doubtful and must be used with great caution.

In biological matters, Gallant draws heavily on G.L. Faber's work from 1883 and his observations on fisheries ecology, especially with regard to the exploitation of pelagic species. This is problematic since marine fisheries ecology was still in its early stages at that time and little was known about the interaction between fishing and fish abundances.[6]

Using Faber in this uncritical way leads Gallant to some erroneous conclusions, e.g., he quotes Faber as saying that fishing has little or no influence on the catch of pelagic species, and that the wide fluctuations between annual catches are only due to natural phenomena such as climate.[7] At the time it was widely accepted that due to massive spawning, the recruitment of juvenile fish could not be affected seriously by fishing.[8] Today we know that this is certainly not the case. In reality the fluctuations in yearly catches observed by Faber could have been caused by overfishing as well as by natural phenomena.

The use of outdated fishery biology led Gallant to conclude – inaccurately – that the irrational movement and fluctuation of pelagic species did not allow them to be commercially exploited in antiquity. Clearly, Gallant's source material is not optimal, and in some cases out of date, but is it at all possible to use modern fishery statistics to estimate the likely output of a historical fishery and the level of exploitation?

2. *Ecosystem changes*

From a biological point of view, Gallant makes the serious mistake of seeing nature as a constant factor, which doesn't change over time and space. In reality, ecosystems change and fluctuate over periods of time. In a marine ecosystem, these fluctuations will affect the abundance of fish and therefore eventually the catches.[9] These fluctuations can originate from natural processes such as temperature, salinity, primary production, predator – prey relationships, etc., but also from human activities such as fishing, draining and pollution.[10]

When Gallant argues that it is possible to compare ancient fishing methods and modern catch data, he also assumes that the marine environment has not changed, that temperature, salinity etc. have always been the same, that the abundance of species is identical, and that human exploitation has not had any

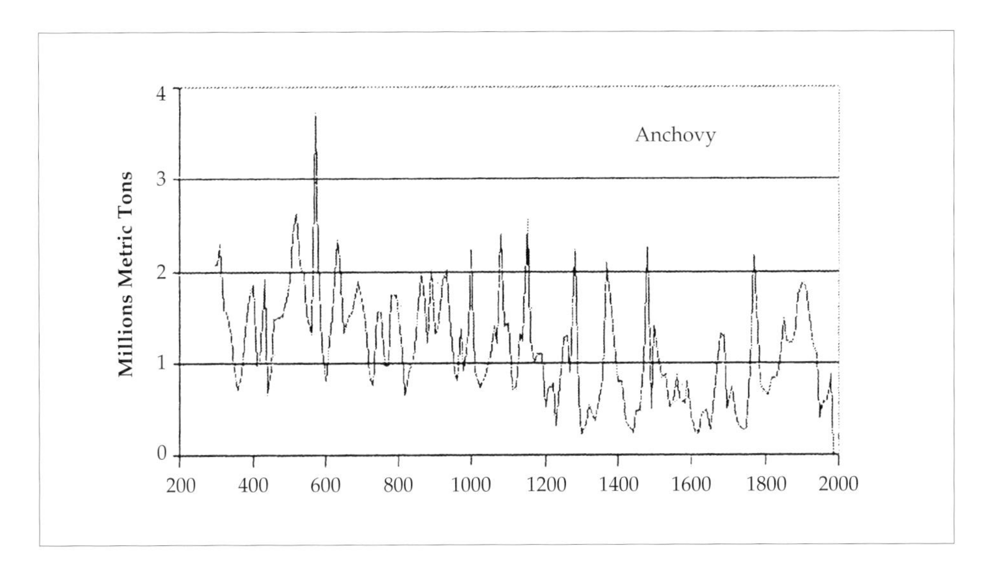

Fig. 1. From R.C. Frances, Exploited seas, *2001, 134.*

affect on the fish stocks. Two examples will show that ecosystems can indeed change dramatically over time, due to both natural and human factors.

2.1. Example 1: Anchovy in the Pacific

The case of the Coastal Pelagic Abundance of Anchovy in the California Current Ecosystem is an example that illustrates nature driven changes in the abundance of fish.

The change in the distribution of the stock seems to be a natural phenomenon, since the anchovy was not commercially exploited before the end of the 19th century. In the case of the Californian anchovy, the climate seems to play a significant role, and it is possible to establish a link between the cadence of the Pacific Decadal Oscillation (PDO)[11] and the abundance of anchovy. Note that there are both low frequency changes in distribution, as well as long-term changes.[12]

2.2. Example 2: Cod in the North Sea

An example of how human activity can affect the abundance of fish is seen on the graphics representation of the decline of cod stock in the North Sea. The Spawning Stock Biomass expresses the well being of the stock but, as seen on the graph, the cod stock in the North Sea has been declining since 1971 due to high levels of exploitation.[13]

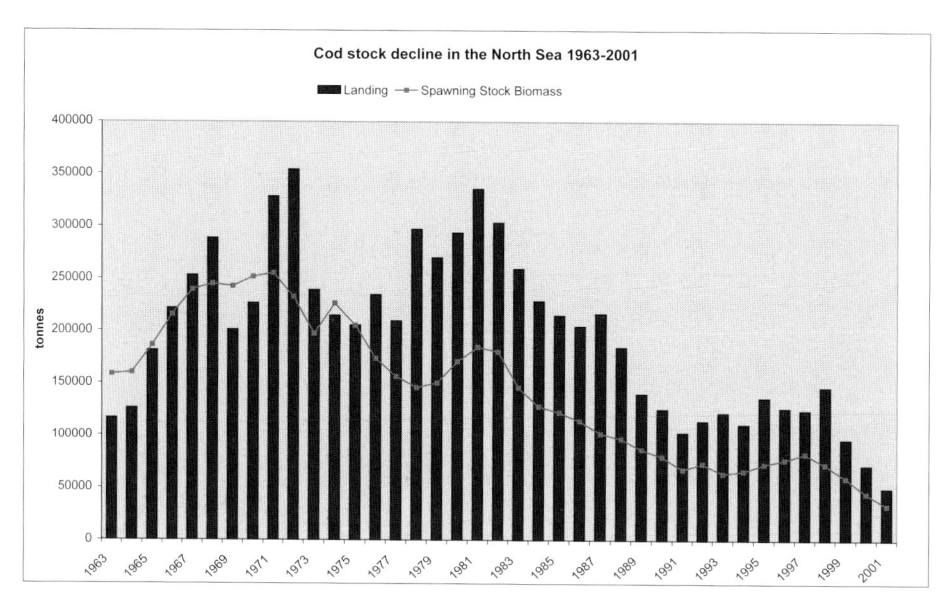

Fig. 2. ACFM Report *2003*[14]

Observe that the catches continue to grow until around 1981, even though, by that date, the stock has been declining over a ten-year period. In a high-efficiency fishery it is possible to maintain high catches even though stock is depleting. As long as the fishing technology becomes correspondingly more efficient, or cheaper, the fisherman's economical income will remain somewhat stable.[15]

When Gallant uses catch statistics from the 1950s and 1960s, we must keep in mind that the fishing effort in the Mediterranean and Black Sea by 1950 and 1960 was enormous compared to the effort in antiquity, and the level of exploitation today is therefore much higher. Thus, one cannot assume that the CPUE is necessarily higher in a high-efficiency fishery than in a more primitive fishery: the modern ecosystem might be depleted, whilst the antique ecosystem may have been at its pristine stage. It is very likely that the main fish stocks and species in the Mediterranean and Black Sea became maximal or over-exploited during the second half of the 20th century, and that the massive fishing effort in the modern period was necessary to maintain an adequate catch.[16] Keeping this in mind, it is quite possible that a *smaller* fishing effort in antiquity would produce a substantially *larger* catch than those of modern times.

The reasons why modern fish-statistics cannot be used to estimate the likely output of ancient fishery can be summarised in two main arguments:

1. That climate changes over time, and climate affects the distribution of species

2. While a large fishing effort today only results in a modest catch due to heavy exploitation of the stocks, a low fishing effort in antiquity could have given a considerable catch.

This means that Gallant's main argument about the ancient fisherman's CPUE being so diminutive compared to modern CPUE might be wrong, because he is comparing two fundamentally different ecosystems.

If one accepts Gallant's data and methodology as somewhat trustworthy, despite the reservations already mentioned, we still need to examine how he interprets the data.

3. Gallant's theses

It is possible to summarise Gallant's main theses about ancient fishery in the Mediterranean and Black Sea in five statements:

1. The fishing technology was too primitive/labour intensive to sustain a large-scale fishery, especially for pelagic species.
2. Fishery statistics from modern times (Mediterranean and Black Sea), where more effective fishing gear was used, indicate that catches in antiquity were much smaller. Modern fishery data from Malaysia where similar fishing technology was used shows that the catch per effort is so low that fishery could only be a part-time occupation, supplementing farming etc.
3. Therefore, fish was fished and eaten locally as a supplement to the daily diet, because the amount of calories gained from the fishery could not feed the fisherman himself.
4. Furthermore the price of [fresh] fish was so high, compared to grain, that it should be considered a luxury food, and therefore fish would not have been an important part of the general diet in antiquity.
5. Finally, large-scale processing and long-distance trade were not possible because the catches were small and irregular and the techniques for preservation inefficient.

This led Gallant to conclude that:

> the role of fishing in the diet and the economy would have been, on the whole, subordinate and supplementary ... Its main function would have been to supply a source of sustinence during periods of food scarcity due to reduced crop yields.[17]

Gallant's first argument about the low catch-per-unit-effort (CPUE) of the ancient fishing gear is of course true compared to more modern gear such as high sea vessels and trawlers. But, in general, he tends to underestimate the efficiency and craftsmanship of the ancient fishing gear. For example he states that the potential effectiveness of gill-nets is very small since they had to be accurately tied and such skills were not sufficiently mastered in antiquity.[18] This contradicts the fact that gill nets are known to have been widely used for fishing in ancient Mesopotamia, long before the existence of industrially manufactured nets and lines.[19]

In the final analysis, the catch power of fishing gear, or potential productivity as Gallant calls it, is not the only thing that determines the size of the catches. If the resources being exploited are plentiful, the CPUE may be high, but if the resource is depleted, the corresponding CPUE will be low. As we have discussed above, the marine ecosystems in antiquity are quite likely to have been healthier than the present ecosystem. From this it follows that Gallant's second thesis is also wrong, since ecosystems fluctuate over time. His argument that the low CPUE in the Malaysian fishery is similar to that in antiquity does not hold, as he is trying to compare two fundamentally different ecosystems that are divided by time and space.

The amount of fish consumed on board or in the fisherman's household is, in general, not included in the official catch statistics, since they are normally based on the actual landings destined for trade.[20] In order to estimate the total harvest from the sea, not only should the consumption of the fisherman and his dependents be added to the catch records, but also the discards at sea of juvenile or commercially worthless species. The catch statistics therefore only cover the marketable surplus, after the needs of the fishermen and their dependents have been met.

Thesis no. 3 is based on the assumption of low output in ancient fishery, but since the output may well have been larger, fish may well have played a more significant role in the diet, and also have been so vast a resource that it might have been a commodity for trade.

Gallant bases his fourth thesis, that fish was a luxury food, on price lists from the Boeotian *polis* of Akraiphia. According to these prices, only the wealthiest citizens could afford to buy fish on a regular basis. However, Gallant overlooks that the prices are probably for fresh fish. Because a fresh fish decays quickly it has to be eaten soon after it is caught—how soon depends on the preferences of the consumers, but it is likely that the fish had to be brought to market and consumed within 1 to 3 days after it was caught.[21] Such a high quality product would obviously be expensive. Furthermore, several of the 12 species Gallant has deduced from the list are only found in salt water, and since Akraiphia is without access to the sea, these fish had to be transported some distance before reaching the consumers, which would add to their cost.

We have to consider that the price list is only for fresh fish brought to market as an article for commercial trade. A lot of fresh fish would have been consumed locally by the fisherman himself or traded as part of a local subsistence economy. So fresh fish might be an expensive imported food item in the polis, but at the same time a common food source consumed locally.

So far, we have only discussed fish as a high quality fresh food item, but it is possible to preserve fish by several methods, thereby extending the time the product can be stored or transported.

In his fifth thesis, Gallant argues that an export oriented fishing industry was not possible because of the fluctuations in catches and the inefficient preservation techniques. His theory about fluctuations in catches is erroneously based on the modern fishery statistics and outdated fish biology, as already discussed, and several studies of processed fish in antiquity indicate that *garum* was an important part of the ancient economy, especially in the Roman Empire.[22]

Another preservation technique, which might prove fruitful in explaining how a significant fishing industry could easily process and trade its products, was the drying of fish. The archaeological evidence for the production of dried fish is non-existent since only a rack or a flat area for drying is needed. Dried fish might be a low quality product, but requires no equipment and can be used both for large and small quantities. This would make it a cheap, non-perishable source of food. It would probably be produced locally and some of it would be part of a local subsistence economy and therefore not visible in the written sources, but some of it might have been traded on a commercial basis to supply the poorest part of the population with cheap proteins.

The lack of archaeological evidence and the low status of the product could be the reason why we do not know much about trade in dried fish.

Conclusion and suggestions for future research

It should by now be evident that estimates of catches and stocks for historical fisheries cannot be based on modern catch data. Instead, estimates must be based on archaeological evidence and historical sources, possibly combined with historical ecology and paleoclimatology.

A systematical mapping of sites where objects related to fishing were found would give some ideas about the distribution of the fishery. Unfortunately, however, fish bones and fishing gear are poorly represented on most sites, but this absence of fishing-related items does not necessarily exclude the occurrence of fishing activities.

One way to approach the question of the extent of the ancient fishery would be to calculate the capacity of the fish processing sites along the Mediterranean and Black Sea coasts. These sites would give an idea about the extent of the garum-related fishery. Another way of addressing the problem would be to use the remains of garum amphorae to estimate the amount of fish used to produce such amounts of garum.

The sites where fresh fish was processed into garum would probably be located near the migration routes of the main pelagic species. If these were mapped according to when they were functioning it might be possible to estimate the historical migration patterns for some of the main pelagic species.

The fishery for fresh fish consumption is hard to estimate as it leaves few archaeological traces, and because some of the catch never reaches the markets but is consumed locally by the fisherman himself. Still, the existence of communities in areas with bad farming land, but with access to marine resources, indicates that for some fishermen fishing was a primary occupation. The question is whether the fisherman only fished for his own consumption, or was fishing for a larger processing industry.

Notes

1 Gallant 1985, 43.
2 Unit-effort can be boats per year, catch per hour, person-days spent fishing etc.
3 Gallant 1985, 12, Fig. 2, Fig 3.
4 *FAO Yearbook* 1950-.
5 Gallant 1985, 24.
6 Smith 1994, 35.
7 E.g., Gallant 1985, 29.
8 Smith 1994, 53-54.
9 Jennings 2001, 88.
10 Coull 1993, 23-27.
11 PDO is a pattern of Pacific climate variability where cold and warm regimes shift in 20-to-30 year cycles.
12 The reconstruction is based on twentieth-century fishery stock assessments and paleo-reconstructions in the Santa Barbara Basin. Francis et al. 2001, 134-135.
13 *ACFM Report* 2003, 35-36.
14 The Advisory Committee on Fishery Management (ACFM) provides advice, on behalf of ICES, on the status of *fish and shellfish stocks* in the Northeast Atlantic Ocean, including the Baltic Sea.
15 Jennings 2001, 227.
16 FAO's *The State of World Fisheries and Aquaculture* 2002, 25-26 states that about 47% of all main marine stocks and species are fully exploited and therefore producing catches that are to (or have reached) the stock's maximum sustainable limits. A further 18% of all marine stocks and species are over-exploited.
17 Gallant 1985, 43-44.
18 Gallant 1985, 21.
19 Sahrhage and Lundbeck, 1992, 35.
20 In the Danish Limfjord fishery in the 1890s a conservative estimate of the value of the fish transferred directly to the fisherman's own household amount to 7% of the total value of the fishery. The Limfjord fishery was market oriented, but in a less commercialized fishery with a lot of semi-professional fishermen the amount could go much higher. Jacobsen, 2003, 14
21 Assuming that no cooling techniques were used.
22 Bekker-Nielsen 2002a, 33.

Fishery in the Life of the Nomadic Population of the Northern Black Sea Area in the Early Iron Age

Nadežda A. Gavriljuk

It is commonly believed that fishing is unusual for traditional nomads as they, being herdsmen, believed fish to be inedible and instead mainly ate meat and drank milk. Consumption of fish by the barbarian population in the northern Black Sea area in the early Iron Age was usually limited geographically to the forest steppe zone. The appearance of fishery in the steppe zone was thought to be connected with either the Greeks of Olbia, Chersonesos, the Bosporan Kingdom, etc. or with the post-Scythian population, for example that of Scythia Minor on the Dnieper from the 2nd century BC.[1] However, archaeological research of recent years, in particular finds from Scythian settlements on both banks of the Dnieper[2] dated to the 4th century BC, as well as the results of the study of the economic history of the steppe Scythians – in particular their domestic production[3] – and an application of general patterns of nomadic economy to the interpretation, allow us to take a fresh view of this problem.

Firstly, it is possible to pinpoint an appreciably earlier appearance of fishery in the steppe zone of Scythia than is normally assumed; secondly, we may confidently connect fish processing with the Scythians among the population of the steppes of the northern Black Sea region; thirdly, it can be demonstrated that not only Eastern civilization, but also the Greek way of life influenced the Scythian fishery. The following sources are available for our reconstruction of nomadic fishery:

1. Results of the reconstruction of the palaeoecological situation in the northern Black Sea region and of nomadic palaeoeconomics.
2. Written sources and ethnographic parallels.
3. Results of field investigations of the material cultures of early nomads and a semantic analysis thereof.

Taken together, all three sources testify to the favourable conditions for the development of fishery in any form. In addition to the utilitarian applicability of nomadic fishery, its cultic aspects may also be considered.

1. *Palaeoenvironment*

There is an extensive river system in the territory of the northern Black Sea steppe zone. In every respect, the organization of a pastoral economy depends on water. Some researchers have constructed a typology of nomadic societies based on whether their settlements were located near water resources or not.[4] Almost all the rivers of the northern Black Sea region had rich natural fresh-water fish resources. The greatest rivers are the Dniester and the *Borysthenes* (Dnieper). Both rivers are especially abundant in fish. The Lower Dnieper (the part of the river *along* and *after* the rapids but before the estuary) traverses the Ukrainian crystal shield. So, in this place the Dnieper provides very clean water, which is necessary for the sturgeons migrating upstream to spawn. A similar situation can be found in the *Hypanis* (Southern Bug), which is another big river West of the Dnieper. The Southern Bug also has its own granite bed and rapids and it was also famous for its shoals of migrating fish. Now both rivers flow together in their estuaries and form the giant combined Dnieper-Bug estuary (today c. 800 km^2) where edible fish are rich in number and variety. The *Borysthenes* (Dnieper) is a typical river of the plain. It features numerous small islands and a system of channels in which fish abounded. Late Medieval sources confirm this.[5] The Greeks had understood these natural conditions as can be gleaned from Herodotos' well-known words about the abundance of fish in Scythia. Herodotos also noted the abundance of salt at the Dnieper's estuary (Hdt. 4.53); salt is of course important for processing fish. Near two major crossings over the river at Chortitsa and Kamenka, which were also important trade routes, there arose a number of hill forts. Kamenka,[6] Kapulovka[7] and Sovutinskoe[8] are the first three sites of Scythian settled life. They appeared in the late 5th to the early 4th centuries BC. Already by the end of the 4th century more than 120 settlements had been established on both banks of the Dnieper.[9]

When the Scythian inhabitants of the dry steppe zone appeared in the northern Black Sea region they were surprised at the sheer number of rivers. Not surprisingly there is a word *mal*, borrowed from Scythian and meaning, "deep stagnant water",[10] which is absent from the terms for waterways in the Ossetian language. However, it is not yet possible to extract much paleo-economic information from such a philological analysis.

Judging the water resources of steppe Scythia in their totality, it can be concluded that the northern Black Sea region in the early Iron Age was a zone abounding in fresh water, providing favourable conditions for the development of fishing based on the fishing of fresh water and migrating salt water fish.

Written sources concerning Scythian fishery are not numerous. Herodotos, who had visited the northern Black Sea region in the early years of the Greek colonization, devoted special attention to the description of the natural resources of the country (4.53, trans. A.D. Godley):

> The fourth is the Borysthenes river. This is the next greatest after the Istros, and the most productive, in our judgment, not only of the Scythian but of all rivers, except the Egyptian Nile, with which no other river can be compared. But of the rest, the Borysthenes is the most productive; it provides the finest and best-nurturing pasture lands for beasts, and the fish in it are beyond all in their excellence and abundance. Its water is most sweet to drink, flowing with a clear current, whereas the other rivers are turbid. There is excellent soil on its banks, and very rich grass where the land is not planted; and self-formed crusts of salt abound at its mouth; it provides great spineless fish, called *antakaios*, for salting, and many other wonderful things besides.

In the opinion of the commentators on Herodotos, the *antakaios* is a fish pertaining to the sturgeon family.[11] This family consists of sturgeon, white sturgeon or beluga (*huso huso,* 17 species of which are known today), and starry sturgeon (*acipenser stellatus*). All these migratory fish are found in the Black and Azov Seas. White sturgeons with a length up to 3 m and a weight of up to 200 kg were well known to the Greeks as edible fish from their native Adriatic Sea.[12] Sturgeons prevailed among fish images on Scythian metal objects.

Another favorite "Scythian" fish type was the catfish. This monster with a length of up to 5 m, weighing up to 300 kg, and found in the Dnieper, was described by Pliny. "In Borysthenes catfish are found of outstanding size without bones or cartilages and with very tasty meat". Meat of these fish was not only tasty, but also had medicinal value. Pliny mentions no fewer than 300 recipes for medicine which included various kinds of catfish (Plin. *HN* 9.45). This fish was also depicted in Scythian tattoos and ornaments (Fig. 1).

2. *Archaeological material, images*

Images of fish were disseminated in the animal style art of the early nomads long before their occurrence in the steppes of the northern Black Sea region. Though rare in the art of the Eurasian nomads, they can be found in the culture of the Altai Scythians. Images of fish decorated felt coverings of saddles from Barrow 1 in the burial grounds of Ak-Alakha 1[13] (Fig. 1.2). The head of a fish was pictured on an object from the 2nd Bašadar Barrow.[14] Images of a fish were among the tattoos on a man's fore-arm from the 2nd Pazyryk Barrow[15] (Fig. 1.1).

Images of fish may be an element that connects the culture of the northern Black Sea region Scythians with the culture of the nomads of the "depths of Asia". In the animal style of the Eurasian nomads the depiction of an eagle tormenting a fish was popular (Fig. 2.8).[16] Ornithologists identify this bird as a *sea eagle*. It is the largest of the Eurasian birds of prey. Sea eagles live near expanses of water and feed on fish. The fish in the depiction is most likely a

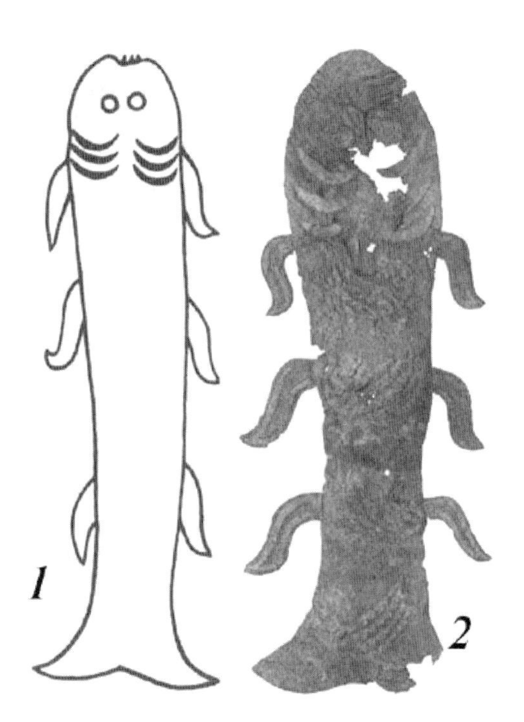

Fig. 1. Images of a sheat-fish. 1: tattoo of the body of the man from the 2nd Pazyryk barrow (by S. Rudenko); 2: decoration of the felt coverings of the saddles from a barrow, No. 1 burial grounds Ak-Alakha 1 (Mountain Altai, 5th century BC) (by N. Polos'mak).

sturgeon. However, the strength of the sea eagle is greatly exaggerated here, as it even has a problem pulling a bream weighing 3-4 kg out of the water. When the bird has seized such a large fish, the sea eagle reaches land by swimming, flapping its wings on the water.[17] These images conceal a double semantic meaning. A fish functions as a symbol of the nether world, while a bird (especially a "regal predator") is a symbol of the upper, heavenly sphere,[18] and in the struggle of the two, the upper world prevails.[19]

The earliest archaeological evidence demonstrating that the Scythians of the northern Black Sea region depicted species found in the underwater world is found on gold plate sheeting covering wooden bowls. Such gold plates were found in Barrow 5 near the village of Archangel'sk in the Cherson region[20] (sea fish, Fig. 2.5), in the Solocha Barrow near Velikaja Znamenka in the Zaporož'e region[21] (river fish, Fig. 2.4), and in the Malaja Simbalka Barrow.[22] All are dated to the end of the 5th century BC. A later development is images of fish on plates decorating horse harnesses. Two silver frontlets (with a gold covering, Fig. 2.6) derive from the Solocha barrow.[23] Bronze ornaments showing sturgeons were found in Barrow 2 near Malaja Lepeticha[24] (Fig. 2.2-3). In the forest steppe zone of the northern Black Sea region a gold frontlet with an image of a fish dated to the 4th century BC – the only such example – was found in Barrow 1 at the village of Volkovcy[25] (Fig. 2.7). Five cast bronze ornaments from a horse's harness with a fish motif were found in the looted Burial 2 of Barrow 1 near the village of Razdol'noe in the Crimea. The find is dated to the 5th century BC.[26]

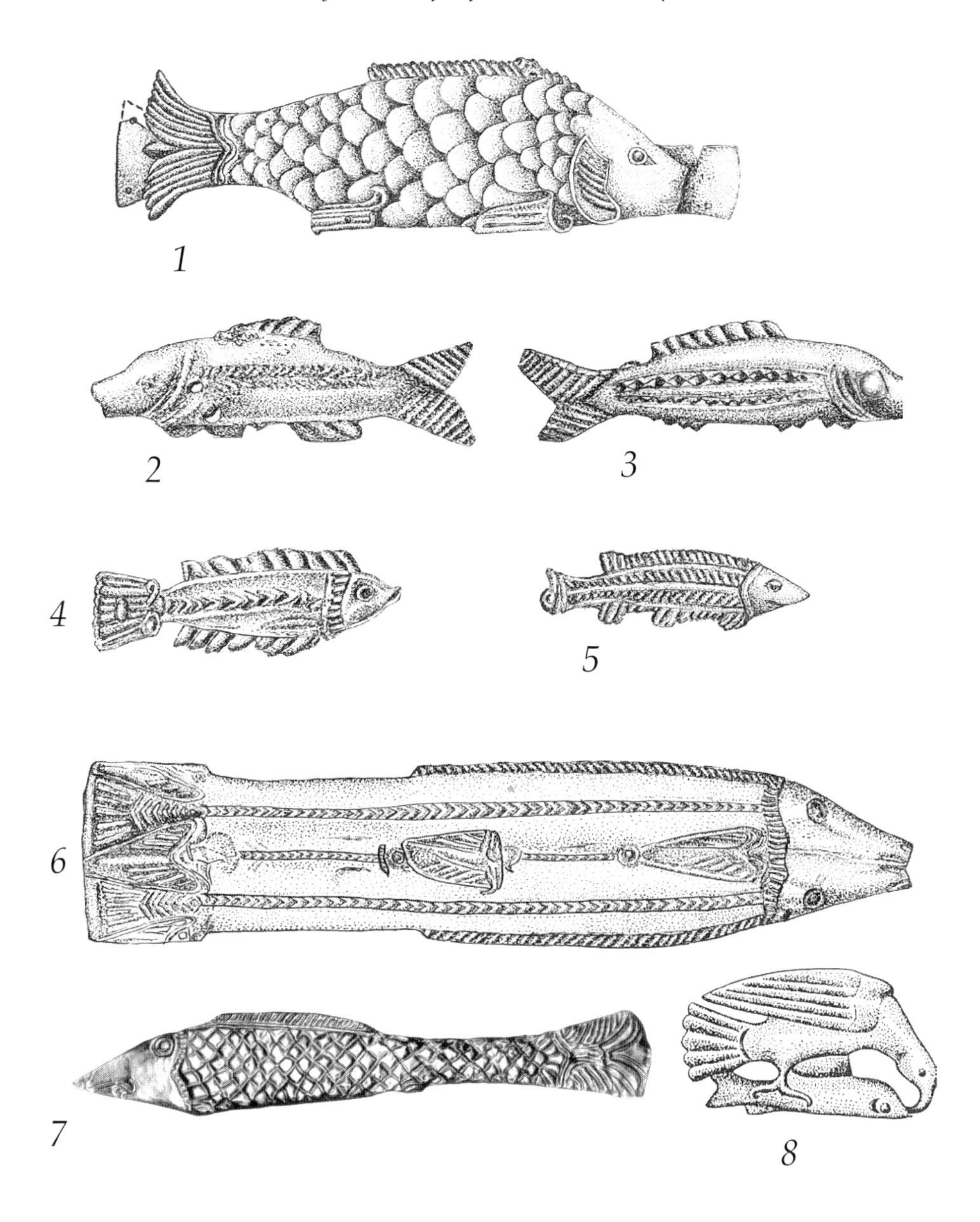

Fig. 2. 1: Decoration in bronze from shield, ca. 400 BC, from Ordžonikidze (Terenožkin, Il'inskaja, Chernenko & Mozolevski 1973, 171), 2-3: Decoration in bronze, from Malaja Lepetikha, 4th century BC (Il'inskaja & Terenožkin 1983, 150, 161), 4-5: Decorations in gold from wooden bowls, late 5th century BC. Fig. 2.4 from the Solocha barrow near Velikaya Znamenka (Mantsevič 1987, 96, N 68), Fig. 2.5 from Archangelsk (Kherson region) (Leskov 1972, 56, fig. 31, 32), 2.6: Gold plated silver frontlet from horse harness, from the Solocha barrow, 4th century BC (Mantsevič 1987, 39-42, N 13,16), 2.7: Gold frontlet from horse harness, from Volkovtsi, 4th century BC (Michel 1995, 217, K3), 2.8: Metal applique (Korol'kova 1998).

The uniquely realistic image of a pikeperch was found in Tomb 1 of the Barrow 12 groups of mine 22 at the village of Ordžonikidze near Dnepropetrovsk (Fig. 2.1). This plate decorated a warrior's shield. The burial dates to the transition from the 5th to the 4th century BC.[27]

Summing up the evidence, in the steppe zone of the Black Sea region the first images of fish are found on wooden bowls (gold sheeting dated to the 5th century BC). In the late 5th to the early 4th century BC fish decorate horse harnesses and even warriors' shields. The fish images probably had an apotropaic character. But in addition to that, the very presence of the ornaments strengthened the protective function of shield and horse harness.

3. *Archaeological material, remains of fish and fishing equipment*

It is not possible to speak about domestic consumption of fish by the steppe Scythians until the 4th century BC. The late 5th to the early 4th century BC witness significant changes in the nomadic economy of the northern Black Sea region, as the economy of the steppe Scythians was transformed from a nomadic to a semi-settled one. Agriculture was practised and part of the nomadic population adopted a settled way of life.[28] The first settlements of steppe Scythians appeared in the Don and Dnieper regions. Finds from these settlements allow us to speak about the occurrence of fish as an element in the diet of their inhabitants. Material from the Elizavetovka and other settlements of the Don estuary testify to the occurrence of fishery in the Scythian steppe zone by the end of the 5th and during the 4th century BC.[29]

Fish bones were found in household pits of the steppe Scythian settlements in the Dnieper area and occasionally in burials of the 5th century BC. In the settlement of Lysaja Gora, two household pits (Nos. 83, 110) were opened which were filled with bones and scales of fish. Fish bones were also found in Pit 1 in the settlement of Černeča. Bones of sturgeon of large and medium sizes predominated.[30]

Reconstruction of the process of fishing based on archaeological material is difficult. However, during the period of spawning the quantity of fish increases so much that "hunting" them with bow and arrow does not seem to have been impossible. Finds of fish hooks in the cultural layer of the Kamenka hill fort allow us to speak about the catching of rather large fish with fishing-lines (Fig. 3.1). Fishing probably also utilised fixed nets, a technique generally considered the oldest.[31]

There is no archaeological evidence for trade in fish on the Dnieper in the 4th century BC. Such trade probably evolved to meet the needs of the Greeks for fish. In the delta of the Don, contacts between Greeks and "barbarians" developed earlier than in the Dnieper region. Therefore "barbarian" settlements with a fishing industry also appeared here earlier, namely in the 4th century BC.[32] Approximately at the same time, fishing on the Dniester also started.[33] The transition to market oriented fishing is often associated with the introduction of large seines and fixed nets.

Fig. 3. 1: Fish hook (iron), 2-4: shuttles for knitting fishing nets (bone), 5-10: sinkers (fragments of amphorae walls).

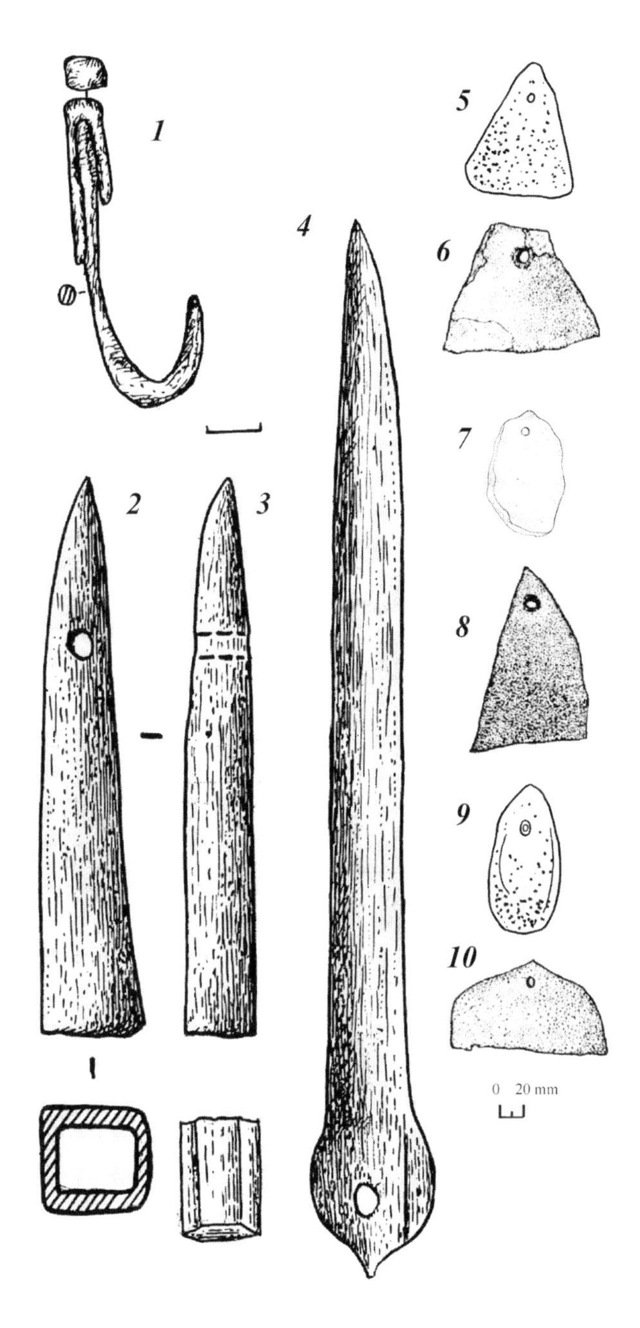

Nets were most likely made of hemp or linen strings.[34] They were apparently knotted with either large or small meshes as seen on a real net found at Nikonion.[35] To ensure the net extended down to the bottom, sinkers were suspended at small intervals along its entire length. In Scythian contexts, sinkers were predominantly made from wall fragments of amphorae, with drilled holes. These have at times been found in excavations of ancient sites. In the Dnieper region such finds appear for the first time in the post-Scythian settlements (Fig. 3.5-10). Among the finds from the settlement of Gavrilovka were plenty of bone objects, among them needles for making or mending fishing nets (Fig. 3.2-3). The abundance of fish bones, finds of sinkers for nets and large needles for making nets at sites dating from the 2nd century BC to the 2nd century AD[36] allow us to speak about a commercial character of fishing by the population in the post-Scythian settlements. Commercial fishing in the Dnieper region, as well as in the Don and Dniester regions, began under the influence of the Greeks.

4. *Conclusions*

1. The nomadic Scythians' familiarity with the underwater world developed long before it became a phenomenon in the steppe zones of the northern Black Sea region and at great distances from the latter area. Fish images are probably an element which connected the culture of the Scythians of the northern Black Sea with the material culture of the nomads from the "depths of Asia".
2. In the steppe zone, the first images of fish were found on the gold sheeting covering wooden vessels dated to the 5th century BC. By the late 5th to the early 4th century BC fish decorate horse harnesses and even shields. Fish images on a horse harness (frontlets and ornaments), not to mention a shield, have both apotropaic and physically protective functions.
3. Archaeological material from sites of the Scythian settled communities that appeared by the late 5th to the early 4th century BC allows us to speak about the occurrence of fish as a component in the diet of the steppe Scythians of the northern Black Sea. For nomads the fishing of freshwater fish was characteristic. This type of fish was a supplementary source of food for the steppe population as some of them went over to a settled life in the 4th century BC.
4. The introduction of market oriented fishing by the population of the steppe zone of the northern Black Sea area was, partly or wholly, in response to the Greek demand for fish.
5. The nomadic population of the Don and Dniester regions began fishing not only for domestic consumption but for trade in the 4th century BC, whereas the "barbarians" and *Mix-Hellenes* of the Lower Dnieper region started "commercial" fishing only in the 2nd century BC.

Notes

1 Grakov 1954, 115; Pogrebova 1958.
2 Grakov 1954, 115; Gavriljuk & Olenkovskij 1992.
3 Gavriljuk 1999.
4 Masanov 1989, 64.
5 Boplan 1990, 45.
6 Grakov 1954, Gavriljuk 1999.

7 Bodjanskij 1951.
8 Ostapenko 2001, 51-68.
9 Gavriljuk & Olenkovskij, 1992.
10 Abaev 1973, 68.
11 Dovatur, Šišova & Kallistov 1982, 282.
12 Lindberg 1971, 50.
13 Polos'mak 2001, 221.
14 Rudenko 1960, pl. CXVI.
15 Rudenko 1953, 136-41.
16 Korol'kova 1998.
17 Gavriljuk, Griščenko & Jablunovskaja-Griščenko 2001. The device of an eagle tormenting a dolphin cannot be "a sketch from nature". No bird of prey is physically capable of dealing with a dolphin in water or lifting it even the smallest distance out of the water. The adult dolphin *(tursiops truncatus)* can weigh up to 300-400 kg, *delphinus delphis* is a little smaller. Even the smallest Black Sea dolphin weighs 10 kg.
18 Tokarev (ed.) 1998.
19 Korol'kova 1998.
20 Leskov 1972, 56, figs. 31, 32.
21 Mancevič 1987, 96, note 68.
22 Alekseev 1995, 54.
23 From N.I. Veselovskij's excavation in 1912: Mancevič 1987, 39-42, notes 13, 16.
24 Il'inskaja & Terenožkin 1983, 150, 161.
25 From S.A. Mazaraki's excavation in 1897-1898: Il'inskaja & Terenožkin 1983, 326, 335.
26 Koltuchov, Kolotuchin & Kislyj 1994, 155.
27 Terenožkin, Il'inskaja, Černenko & Mozolevskij 1973, 171.
28 Gavriljuk 1999.
29 Žitnikov 1992, 72.
30 Gavriljuk, Bylkova & Kravčenko 1992, 22, 24, 75.
31 Zelenin 1991.
32 Žitnikov 1992, 72.
33 Brujako 1999.
34 Zelenin 1991, 106.
35 Brujako 1999.
36 Pogrebova 1958.

Fish and Money: Numismatic Evidence for Black Sea Fishing

Vladimir F. Stolba

Human beings – both ancient and modern – have not only associated the word "fish" with food, but also, to a very great degree, with a marketable commodity linked with money. For such a sea-oriented people as the Greeks, who established settlements on the shores of the Mediterranean and Pontos, and whose dependence on the waterways and marine resources was extraordinary, this association must have been particularly strong. We may assume, therefore, that it was not an inverse association. The sporadic appearance of fish on coins, or as a coin type all around the Greek world, would also suggest that we are not dealing with a fortuitous phenomenon.

In terms of the ancient Black Sea, where the written sources on economic conditions in general – and fishery in particular – are often very scarce, the numismatic data may provide an additional piece of evidence. It is generally accepted that in Archaic and Classical times the typology of the Greek coins was chiefly of a religious character, which it maintained right into the early Hellenistic period. Despite this fact there is a fairly large group of types related one way or another to the local resources that secured a reputation or prosperity for the specific city or entire region.[1] Two of the many examples are the barley ear and barley seeds on the well-known silver specimens of Metapontos and Leontinoi, respectively (Fig. 1.1-2). The grain ear on the fourth-century BC gold staters of Pantikapaion and the wheat seed on the contemporary coins of Phanagoria (Fig. 1.3-4), are also totally consistent with what we learn from Athenian orators (Dem. 20.31-33; Din. 1.43; Isocr. 17.57) about the role of the Bosporos in the international grain trade. An appeal to marine resources was no exception here, and the ubiquitous tunny-fish on the electrum and silver of Kyzikos may serve as an example (Fig. 1.5,11). Perhaps even more explicitly this occurs in the coinage of Gela in southern Sicily where we find a young male head surrounded by fishes representing a local river god (Fig. 1.9).

On the other hand, some emblems, which at first glance seem to belong to the same group, should perhaps not be regarded as such. For instance, the eagle-on-the dolphin symbol occurring on the coins of Sinope, Istros, and Olbia (Fig. 1.6-8) can hardly be seen as an allusion to the marine resources of these cities, but perhaps to their coastal position.[2]

More than twenty years ago P.O. Karyškovskij, who discussed this issue at length, was inclined to see the dolphin and the eagle as attributes of Apollo

Fig. 1. Greek coins of the Classical and Hellenistic periods. 1) Metapontos, AR; 2) Leontinoi, AR; 3) Pantikapaion, AU; 4) Phanagoria, AR; 5) Kyzikos, EL; 6) Olbia, AE; 7) Sinope, AR; 8) Histria, AR; 9) Gela, AR; 10) Akragas, AR; 11) Kyzikos, AR. (1-2, 11: Gorny & Mosch auctions (126, lot No. 1101; 122, lot No. 1099; 121, lot No. 164), photo courtesy of the Gorny & Mosch Giessener Münzhandlung; 4: in commerce; 3, 6-8: Danish National Museum, Collection of Coins and Medals, SNG Cop. 6.20, 6.75, 18.281, 6.191, photo courtesy of the Museum; 5: CNG auction 14.01.2003, lot No. 301, photo courtesy of the Classical Numismatic Group, Inc.; 9-10: after Kraay 1976, pls. 48.826 and 46.797).

Delphinios and of Zeus, respectively.[3] But semantically speaking, the question is rather intricate, since a similar design, sometimes with a fish instead of a dolphin, occurs both in the Scythian and Graeco-Scythian art, and was also distributed far beyond areas of Greek culture as such. Perhaps it should be understood as one of the main cosmological symbols of the ancient inhabitants of Eurasia, where the eagle seemingly represents the celestial or the upper-world whereas the dolphin/fish could represent the water, i.e. the *under*world.[4] Thus, for the Indo-Europeans a combination of the two could mean, as suggested by some scholars, a sacred marriage alliance between the sky, as a male substance, and the terrestrial or aquatic element, as a female substance, something that ultimately guaranteed the existence of everything.[5] Describing the universe by means of a zoological code might though imply both the unity and the conflict of opposing principles. In this sense it is to some extent semantically similar to the well-known scenes of a wild beast attacking a herbivore.[6] Indeed, the Greeks might interpret this notion in a slightly different way, correlating these two elements with the sacred images with which they were more familiar. To illustrate this explanation, one might refer to the numismatic parallel from south-western Sicily, namely the coinage of the non-coastal *polis* of Akragas. Not long before its destruction by the Carthaginians, the city struck very elegant dekadrachms, the design of which was apparently intended to celebrate the Olympic victory of the Akragantine Exainetos in 412 BC.[7] On the obverse of these coins, the chariot of Helios runs between the sky and the sea, which are represented by an eagle and a crab, correspondingly (Fig. 1.10).

In order to avoid any further confusion, however, we shall concentrate henceforth only on the fish, leaving aside numerous representations of dolphins. The evidence is organised geographically starting from the north-western corner of the Black Sea and following its shores clock-wise.

1. *Karkinitis*

Karkinitian coins (Fig. 2.1-3) revealing a fish as a main coin type are not numerous. It is not long ago, that as a result of excavations of 1980 to 1982 in Eupatoria, they were introduced to the scientific world.[8] All the specimens are bronze and made in the cast technique. This peculiarity strongly indicates the influence from the neighbouring city of Olbia, where this distinctive technique, foreign to the Greek world as such, was employed from the sixth century BC onwards. Archaeological context and parallels in the numismatics of Olbia date the issues reliably to the early fifth century BC. According to shape they may be divided into two main groups.

The figured cast specimens in the shape of a fish constitute the first of these groups. In fact, only one side of the casts represents the fish in relief, while the other having a long horizontal rib resembles rather an arrowhead (Fig. 2.1).

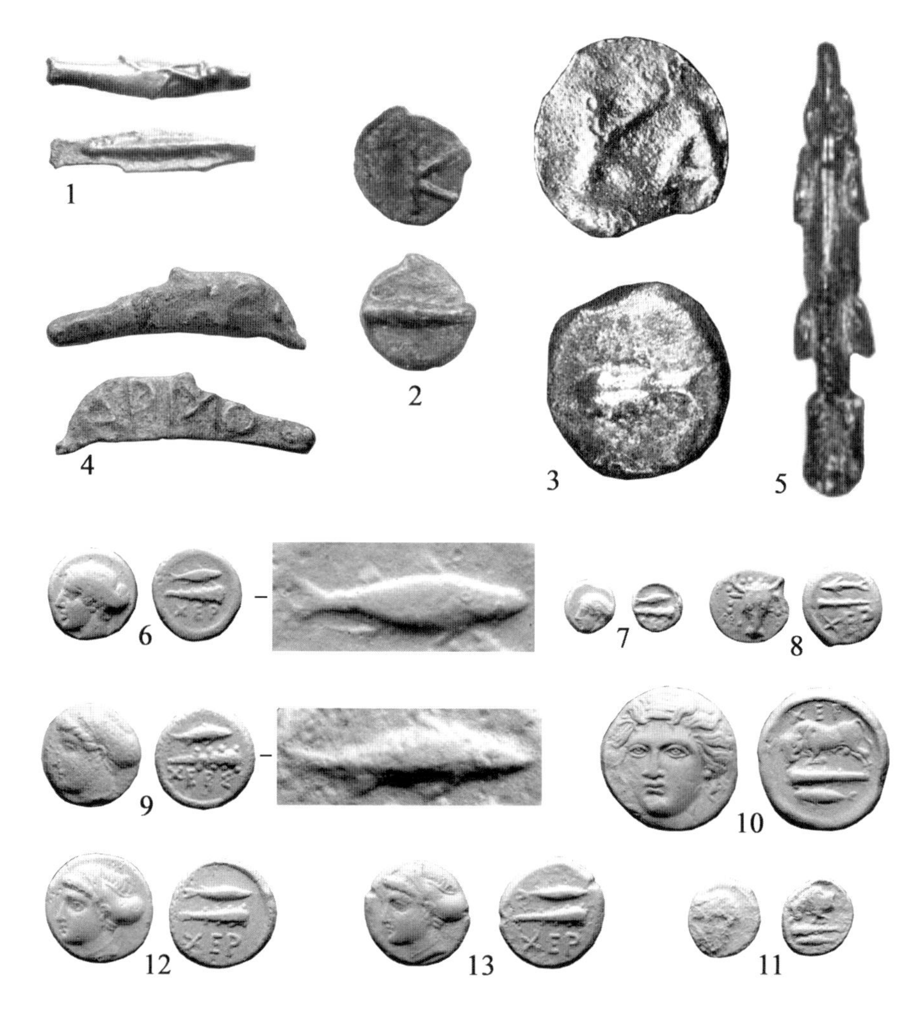

Fig. 2. Coins of Karkinitis, Olbia and Chersonesos. 1-3) Karkinitis, AE; 4) Olbia, AE; 5) Sturgeon shaped bronze figure from barrow 4 near the village of Ryleevka (West Crimea); 6-7, 10-11) Chersonesos, AR; 8-9, 12-13) Chersonesos, AE. (1: Gorny & Mosch auction 60, lot No. 180, photo courtesy of the Gorny & Mosch Giessener Münzhandlung; 2: Odessa Museum of Numismatics, photo courtesy of the Museum; 3: after Kutajsov 1986, fig. 1; 5: after Koltuchov 1997, 63, fig. 3; 6-8, 11: State Hermitage Museum, Numismatic Department, inv.-nos. 25936-25937, 26075, 25945, after casts; 9: Bibliothèque royal de Belgique, Cabinet des Médailles, L. de Hirsch Collection 850, after a cast; 10: Hess-Leu auction 2.04.1958, lot No. 119, after a cast; 12: Ashmolean Museum Oxford, Heberden Coin Room, May bequest 1961, after a cast; 13: Bibliothèque National Paris, Cabinet des Medailles, after a cast.)

The second group is round in shape and consists of two denominations showing a fish on the obverse and an abbreviated city-ethnic KA or K on the reverse (Fig. 2.2-3).

Kutajsov, who first published and attributed these coins to Karkinitis, considered the obverse of the last group to be a representation of a dolphin or, as he suggested later, one of the sturgeon types. However, taking into account their state of preservation and the rather careless execution of the moulds, any attempt to identify the fish species should be met with caution. The dolphin, it seems, has the least chance of being among the candidates here.

Indeed, more helpful in this respect is the first of the two issues. The general outline of the casts, the heterocercal caudal fin with its characteristically elongated upper part and slightly upturned snout, leave little doubt that we have a representation of one of the sturgeon species, as already recognised by the first publishers. A relief horizontal rib, discernible on some of the casts, might perhaps also be regarded as a poor reproduction of a scute row, which distinguishes this kind of fish. However, the outward appearance showing the features characteristic for the entire *Acipenseridae* (Sturgeons) family is not as detailed as to make one agree unreservedly with M. Zolotarev, who identifies it as the *Huso huso* or beluga. As suggested by S.G. Koltuchov, the peculiar form of the Karkinitian cast money could perhaps have affected the appearance of the sturgeon in the contemporary Scythian animal-style metalwork.[9] Articles designed or shaped like fish are fairly widespread in the steppes of Southern Russia showing an evident concentration in the Lower Dnieper region.[10] Recent finds from the barrows near the village Ryleevka in the north-western Crimea may provide one of the most characteristic examples of such representations (Fig. 2.5).[11] It is, however, beyond the scope of this article to become involved in the details of this issue, although, as in the case of the arrowhead money, we certainly cannot exclude the cultural influence from Scythia upon the Greeks, rather than vice versa.

2. *Chersonesos*

2.1 Obv. Parthenos, left.
Rev. Fish r., and club, XEP. AR and Æ.
Anochin 1977, nos. 1-7; *SNG BM* 706.

This type is represented in two metals, which apparently were struck contemporaneously. Well-preserved specimens of two different denominations kept in St. Petersburg, London, and Oxford[12] allow more accurate attribution of the fish species (Fig. 2.6,12). Two clearly discernible dorsal fins and a projecting anal fin seem to indicate that the die engraver intended to represent a mullet. Perhaps this might not be true, however, for the variety of this type with the one-letter longer legend XEPΣ on the reverse, the finest example of this being on the coin from the L. de Hirsch collection in Brussels (Fig. 2.9).[13] Despite the exceptional state of preservation only one dorsal fin can be identified on this specimen with certainty. If this is the case, the most likely candidate would indeed be a Pontic shad.

> 2.2 Obv. Bukranion.
> Rev. Fish left or right, and a club beneath. XEP. Æ.
> Anochin 1977, Nos. 9-12.

The type is represented by bronze specimens only (Fig. 2.8). In all dies known to me, the fish has apparently only one dorsal fin, although the entire image is so stylised that any attempt to identify the species would be a matter of pure speculation. The possibility cannot even be ruled out that we are dealing with an extremely poor representation of a dolphin, such as that appearing not infrequently on the coinage of Classical and early Hellenistic Byzantion.[14]

While the silver and bronze of Type 1 belong to the first quarter of the fourth century, Type 2 known only in bronze cannot be dated earlier than the second quarter of the same century. Amazing though it may seem, apart from the club the early coin typology of Chersonesos has little to do with that of the metropolis. Permanence of the reverse device, which characterises the local coinage for at least a quarter a century, seems to imply an effort to introduce it as an emblem of the city. It can perhaps be corroborated by the following type:

> 2.3 Obv. Head of Parthenos in a three-quarter view.
> Rev. Butting bull, l.; club and fish beneath. AR.
> Anochin 1977, Nos. 23-25.

The same emblem appears here as an additional element of the type (Fig. 2.10-11). However, this attempt seemed to fail and from about 360 BC, the fish, unlike the club, disappears entirely from the Chersonesean coin typology. Although the reason for this alteration remains unknown, we may assume that the annexation of the fertile plain of the western Crimea, where the earliest Chersonesean presence recorded at Panskoye I is datable to exactly the same period,[15] could perhaps lead to the shifting accents in the *polis'* economy.

In spite of this it would be erroneous to think that fishing was losing its importance in the following periods of the city's history. On the contrary, fish was certainly both staple food and one of the bases of the Chersonesean economy,[16] and it maintains this role in the modern city of Sevastopol', the successor to ancient Chersonesos. Perhaps this is most vividly stressed in the novel *Lestrigonoi* by the early twentieth century Russian writer Alexander Kuprin, who gives an account of the everyday life of the fishing village of Balaklava just on the outskirts of Sevastopol.[17] This story makes clear the role of the dolphins too, which chase the huge schools of mullet into the deep – but extremely narrow – Balaklava Bay, thereby providing a unique opportunity to catch the fish in enormous quantities.[18] Characteristically, the images of dolphins occur occasionally on the Chersonesean small copper coins at that very point when we find the depiction of a fish. However, taking into account bone remains of the common dolphin (*Delphinus delphis ponticus* Barab.-Nik.) reported from the rural settlements of the western Crimea[19] and Chersonesos itself,[20] it would seem probable that in some periods it might have been hunted for its meat or oil as well.

3. *Pantikapaion*

3.1. Obv. Lion's head facing.
Rev. Ram's head l.; below, sturgeon l.; ΠΑΝΤΙ. AR.
Anochin 1986, Nos. 67-69; *SNG BM* 852-853.

3.2 Obv. Head of a bearded satyr right.
Rev. Forepart of an eagle-headed griffin left,
underneath a sturgeon l.; ΠΑΝ. Æ.
Anochin 1986, No. 111; *SNG BM* 869-871.

3.3 Obv. Head of a bearded satyr wearing a wreath, left.
Rev. Head of a sturgeon r.; ΠΑΝ. Æ.
Anochin 1986, No. 81.

3.4 Obv. Beardless head of satyr with wreath left.
Rev. Head of a lion, l., with a sturgeon beneath it; ΠΑΝ. Æ.
Anochin 1986, No. 125; *SNG BM* 883-885.

Apart from silver coins of Type 1, dating to the late fifth century BC (Fig. 3.1-2), the remaining coins belong to the late fourth century BC and are bronze. The issue of the last type was particularly abundant, and the period of circulation fairly long.

Elements of the types are detailed enough to make it certain that they represent one and the same fish species. However, as to its attribution, opinions are not unanimous. Zograph sometimes calls it, "fish of the sturgeon family", sometimes sterlet.[21] D.B. Šelov was inclined to see here a Russian sturgeon.[22] Considering it to be the same sort of reflection of local conditions as the horse's head, bull's head, the ear of grain, he wrote: "this image ... unquestionably points toward Bosporos' wealth in fish and the importance of the fishery industry for the country's economy".[23]

As noticed already by a number of scholars, the composition of the entire Type 3 with a lion's head to the left in the centre, the letters in field to the sides, and with a fish beneath, clearly reminds one of the reverses of the tetradrachms struck in Kyzikos in the fourth century BC,[24] which perhaps served as originals for the local copper currency (Fig. 1.11).

Fig. 3. Coins of Pantikapaion. 1-2) AR; 3-5) AE. Sturgeon species: a) Beluga; b) Russian sturgeon; c) Starry sturgeon (sevryuga); d) Fringebarbel sturgeon; e) Sterlet. (1-2: after Anochin 1986, nos. 67-68; 3: Museum Narodowe Warsaw, inv.-No. 105512, after a cast; 4: Gorny & Mosch auction 118, lot No. 1150, photo courtesy of the Gorny & Mosch Giessener Münzhandlung; 5: Danish National Museum, Collection of Coins and Medals, SNG Cop. 6.35, photo courtesy of the Museum.)

In 1964 this motif was discussed in a special article by V.M. Brabič. Following Šelov in identifying the fish as a Russian sturgeon, he suggested regarding the entire composition as semantically interdependent. According to this view, both lion and griffin appear to carry out a protective function regarding grain and fish, which were the basic commodities of the Bosporan trade.[25] Taking into account the Greek belief that the griffins guarded gold from the Arimaspians on the northern edge of *oikoumene* (Hdt. 3.116.1; 4.13.1. Cf. Aisch., *Pers.* 804) this cannot be completely ruled out. However, it seems that the coin emblems allow more accurate attribution of the fish species. A distinctive long snout pointed at the tip, which is clearly discernible on the well-preserved specimens (Fig. 3.3-5), as well as the rather narrow body, speak in favour of a sevriuga, known also as the starry sturgeon (*Acipenser stellatus*). This species is probably intended by Athenaios when, speaking about Bosporan sturgeons, he mentions *genos oxyrinchos* or the sharp-snouted variety as "not inglorious in the eye of mortals" (Athen. *Deipn.* 3.116b).

The above is consistent with the fact that among sturgeons it is precisely the sevriuga which prevails in the icthyo-faunal remains from the Bosporan sites. So in the finds from Pantikapaion, analysed by V.D. Lebedev and Ju.E. Lapin, the sevriuga makes up 12.8% against 10.2% and 7.7% for Russian sturgeon and sterlet, respectively.[26] In the finds from Phanagoria the share of the sevriuga was 30.8%, 22.2% being Russian sturgeon.[27] *Acipenser stellatus* equally predominates in the modern catches in the Kuban River,[28] which in antiquity had its main out-fall not in the Sea of Azov as today, but further south in the Black Sea near the straits or even in the Taman Bay. Bearing this in mind we may also assume here some higher concentration of sturgeons than in our time.

Indicating the sources of the economic prosperity of Pantikapaion, the fish on its coins might well have had a double significance, implying at the same time the city-name. Being related to Pantikapes, one of the main Scythian rivers mentioned by Herodotos (4.54), it apparently derives, according to M. Vasmer and V.I. Abaev, from the Old-Iranian **panti-kāpa*, which should mean a "fishy way".[29]

Strabon (7.3.18) provides additional evidence on the Bosporan fishery while describing the winter extremes of the region. "The severity of the frost" – he says – "is most clearly evidenced by what takes place in the region of the mouth of Lake Maeotis: the waterway from Pantikapaion across to Phanagoria is traversed by wagons, so that it is both ice and roadway. And fish that become caught in the ice are obtained by digging with an implement called *gangame* and particularly the *antacaei*, which are about the size of dolphins" (transl. H.L. Jones). H.F. Tozer supposed here that it was ice fishing by means of a small round net which is denoted by the term *gangame*.[30] In contrast, V.Ju. Marti and H.L. Jones commenting on the same passage assume it indicates a different technique.[31] "Strabo", Jones wrote, "seems to mean that the fish were embedded in the ice",[32] while the *gangame* refers rather to a pronged instrument resembling a trident.[33] This assumption, however, appears in both respects to be a matter of confusion. According to Oppian and lexicographers *gangame* is a variety or synonym for *sagene* and *diktyon*, both of which mean "fishing net".[34] Both A.W. Mair and F. Fajen, the translators of Oppian's *Halieutika*, see it as a "drag-net" or "Schleppnetz".[35] Moreover, the change of climatic conditions since the Late Classical period was insignificant,[36] and assuming even the severest frost which might occur in the region, the thought that the fish could be frozen into the ice, is rather dubious. Furthermore, in Strabo's account we face another difficulty, when in connection with ice fishing he mentions sturgeons. Although the catching of sturgeons could in principle take place in winter, these species seem not to lend themselves to ice fishing. In winter while hibernating they congregate in sea-bottom holes and exhibit little activity. In spring, when the ice breaks, they rise from the bottom holes and move upstream for spawning.

The next group of coin types showing fish leads us to the southern coast of the Black Sea. The first area is represented by Sinope and Herakleia. Despite the long history of their coinage, which goes back to the sixth century and the last quarter of the fifth century BC, respectively, a fish appears solely on a few types of bronze from the imperial time. The quality of the images does not allow any reliable identification of the fish species.

4. *Sinope*

Caracalla

4.1 Obv. Bearded head right; ANTONINVS AVG.
Rev. Fish left; C I F SINOP. Æ.
Rec. I.1, 205, No. 134, pl. 28.4.

Geta

4.2 Obv. Head of Geta right; IMP SEPTI GETA.
Rev. Fish left; C I F SI NOPES. Æ.
Rec. I.1, 206, No. 141.

4.3 Obv. Head of Geta right; C P SEPT GETA.
Rev. Fish right; C·I·F· SINOPES. Æ.
Rec. I.1, 206, No. 142, pl. 28.10; *SNG Cop.* 317.

Alexander Severus

4.4 Obv. Bust of the emperor right; AV·SEV·AΛEXAND.
Rev. Fish left; [C·I·F]S·A·CCXC·III[I?]. Æ.
Rec. I.1, 207, No. 148.

Maximus

4.5 Obv. Bust right; MAXIMVS CAES.
Rev. Fish left; C·R·I·F·S·A[---].
Rec. I.1, 207, No. 153, pl. 28.18.

5. *Herakleia*

Obv. Herm of Dionysos; ΗΡΑΚΛΕΩΤΑΝ.
Rev. Two tunny fish to l. and r.; in centre, pellet. Æ.
SNG BM 1639.

Fig. 4. Bronze coins of the southern Black Sea littoral. 1) Sinope, Caracalla; 2) Sinope, Geta; 3) Sinope, Maximus; 4) Herakleia Pontike, enlarged 1:1.5; 5) Byzantion, Caligula; 6-7) Byzantion, Plotina; 8) Byzantion, Sabina; 9) Byzantion, Faustina the Younger; 10) Byzantion, Lucilla. (1-3: after Waddington, Babelon & Reinach 1904, pl. 28.4, 10, 18; 4: British Museum, Department of Coins and Medals, SNG BM 1639, photo courtesy of the Museum (Andrew Meadows); 5-10: after Schönert-Geiss 1972, pls. 63.1312/2, 68.1361-1362/2, 69.1374, 73.1420/2, 74.1422/2.)

In the *Sylloge* of the British Museum the Herakleian type is dated (very) approximately from the early second to the late first centuries BC. However, I cannot see any reason for such an early date. On the contrary, taking into account the form of the letters as well as the specific design of the reverse type paralleled in the coinage of Byzantion, it is more likely that we are dealing with a so-called pseudo-autonomous issue of the Late Roman period.

6. *Byzantion*

The coinage of Byzantion offers us further examples of types representing fish, although, to be more precise, we are talking about one and the same reverse emblem reproduced repeatedly over more than two hundred years. Apart from minor variations the composition constituted by two tunny fish does not show much diversity. On the earliest specimens struck in the name of Caligula, Trajan, and Sabina the fishes appear alone and, as a rule, facing in the same direction. However, the coins of Plotina, the wife of Trajan, already reveal further development of the type by adding a dolphin between the fish. In this form it survives until the middle of the third century. Starting from Plotina, we see the two tunny fish regularly turned in opposite directions.

Concerning this type, E. Schönert-Geiss in her *Corpus* of the coins of Byzantion of the period of the Roman Empire wrote: „Die Thunfischerei scheint auch in römischer Zeit noch immer mit zu der wichtigsten Einnahmequelle der Stadt gehört zu haben. Das lässt sich jedenfalls an den zahlreichen Abbildungen zweier Thunfische – dazwischen häufig ein Delphin als zusätzliches Symbol für das Meer – erkennen."[37]

This assumption is completely consistent with the remark by Athenaios when he says that the Byzantians "have so many fish in their part of the world that they are all clammy and full of phlegm" (Athen. 4.132e). As to a description of the city, Polybios' words are even more precise: "the site of Byzantion is as regards the sea more favourable to security and prosperity than that of any other city in the world known to us, but as regards the land it is most disadvantageous in both respects" (Polyb. 4.38.1).

Taking into account the above mentioned, it is tempting to lean towards the statement of Schönert-Geiss. However, it turns out that the type being discussed seems to have very little if anything to do with the fishing industry of the *polis*. Being mostly religiously or mythologically determined, the coin types reveal no connections with any of the city's economic activities. Furthermore, the fish is well known as an emblem of the Syrian Goddess, Atargatis. The fish is one of the elements of her cult legends and in some respects her physical appearance was that of a fish (Lukianos, *On the Syrian Goddess* 14).[38] The cult of Syrian Aphrodite and *Dea Syria* seems rather early to spread to the various parts of the Greek world where she was generally regarded as Syrian Aphrodite. The dedications from Berezan,[39] Olbia[40] and Bizone[41] prove that her cult reached as far north as the Ukrainian and Bulgar-

Fig. 5. Bronze coins of Anchialos. 1) Crispina; 2) Julia Domna; 3-5) Maximinus; 6) Gordianus III. (1: after Struck 1912, pl. 6.22; 2: auction Gorny & Mosch 118, lot No. 1631, photo courtesy of the Gorny & Mosch Giessener Münzhandlung; 3: photo courtesy of the Aeqvitas.com (Heather Howard); 4: photo courtesy of Thomas Burger; 5: auction Lanz 102, lot No. 831, photo courtesy of the Numismatik Lanz; 6: in commerce, photo courtesy of the Classical Numismatic Group, Inc.)

ian coasts of the Black Sea.[42] Being regarded as a goddess of fertility she was particularly popular among the female population. Apparently therefore, it is not fortuitous that the overwhelming majority of the coin types of the city showing two fishes were issued in the name of empresses, while the emperors mostly preferred other emblems.[43]

The same is true for the bronze coins of Anchialos struck in the name of Faustina Junior, Crispina, Julia Domna, Maximinus Thrax and Gordian III, which conclude my catalogue.[44]

7. *Anchialos*

Faustina Junior

7.1 Obv. Head of Faustina right; ΦΑΥCΤΕΙΝΑ CΕΒΑCΤΗ.
Rev. Dolphin between two fish; ΑΓΧΙΑΛΕΩΝ. Æ.
AMNG 435; Mušmov 1912, No. 2788, pl. 17.8.

Crispina

7.2 Obv. Head of Crispina right; ΚΡΙСΠΕΙΝΑ СΕΒΑСΤΗ.
Rev. Bigger fish r. between two smaller fish l.; ΑΝΧΙΑΛΕΩΝ. Æ.
AMNG 453, pl. 6.22; *SNG Cop.* 431.

Julia Domna

7.3a Obv. Head of Julia Domna right; ΙΟΥΛΙΑ ΔΟΜΝΑ СΕΒ.
Rev. Dolphin between two fish, in field Γ; ΑΓΧΙΑΛΕΩΝ. Æ.
AMNG 507-508, pl. 7.5; Mušmov 1912, No. 2841, pl. 20.10.

7.3b Obv. Head of Julia Domna right; ΙΟΥΛΙΑ ΔΟΜΝΑ СΕΒ.
Rev. Bigger fish r. between two smaller fishes l.; ΑΓΧΙΑΛΕΩΝ. Æ.
AMNG 509.

Maximinus

7.4 Obv. Laureate head right; ΑΥΤ ΜΑΞΙΜΕΙΝΟС ΕΥСΕΒΗС ΑΥΓ.
Rev. Dolphin between two fish. ΑΓΧΙΑΛΕΩΝ. Æ.
AMNG 604-605, pl. 7.38; Mušmov 1912, 2893.

Gordian III

7.5 Obv. Laureate head of Gordian right; ΑΥΤ Κ Μ ΑΝΤ ΓΟΡΔΙΑΝΟΣ ΑΥΓ.
Rev. Dolphin between two fish. Æ.
AMNG 645; Mušmov 1912, 2923, pl. 17.8.

The resemblance of their reverse type to that of Byzantion is so striking as to conclude there was direct adoption from the latter city.[45]

8. *Conclusions*

Summing up, we may assert that in a number of cases the coin typology of the Greek cities around the Black Sea reflects their dependency on the marine resources both in terms of daily food supply and international trade. However, as we could see, the distribution of evidence is not homogeneous, neither in geographical nor in chronological respects. This does not mean of course that fishery was necessarily of less – or of no – importance for areas and periods which do not match our list.[46] This might have occurred when the development of the local coin types had been determined by different reasons, such as religion, politics, or others, which as yet escape us.

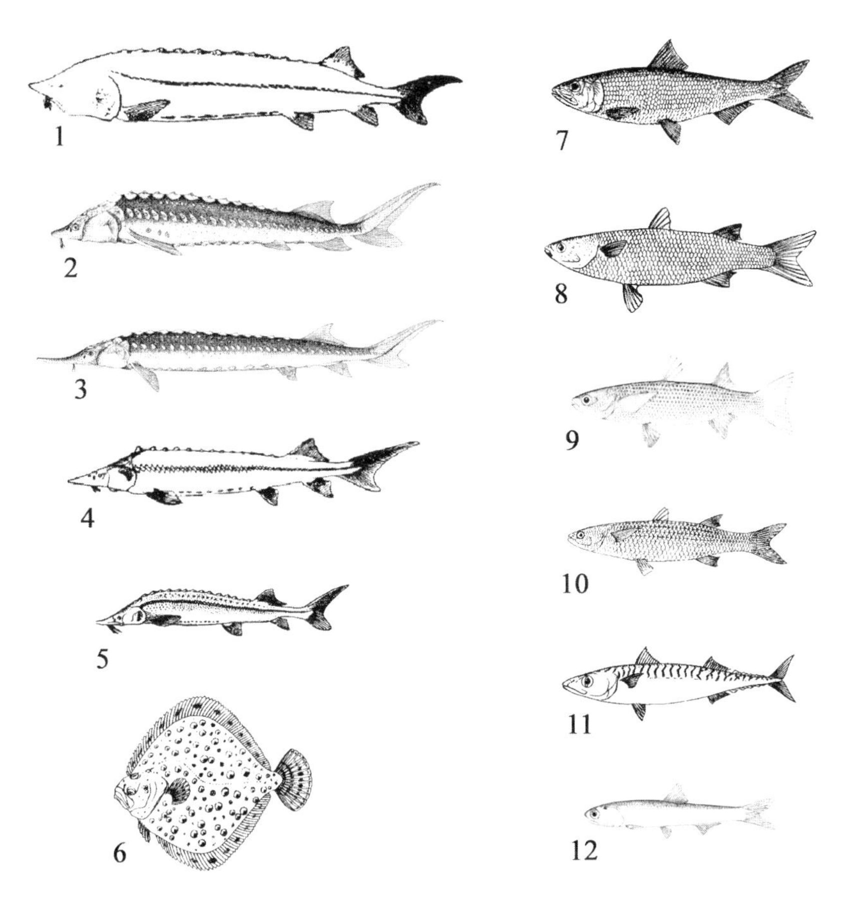

Fig. 6. Main Black Sea fish species of commercial importance. 1) Beluga, Huso huso; 2) Russian sturgeon, Acipenser gueldenstaedtii; 3) Starry sturgeon/ sevryuga, Acipenser stellatus; 4) Fringebarbel sturgeon, Acipenser nudiventris; 5) Sterlet, Acipenser ruthenus; 6) Turbot, Rhombus maeoticus; 7) Pontic shad, Alosa pontica; 8) Flathead mullet, Mugil cephalus; 9) Golden grey mullet, Liza aurata; 10) Leaping mullet, Liza saliens; 11) Atlantic mackerel, Scomber scombrus; 12) Black Sea anchovy, Engraulis encrasicolus. (1, 4-8, 10-11: after http://www.internevod.com/rus/academy/bio/opr; 2-3, 9, 12: after http://www.fishbase.org).

On the other hand, we do not see a great diversity among the species appearing as a coin type or as part of one. Apart from a few cases we must be cautious, however, about inferring that one kind of fish was more important than another. While interpreting coin evidence it has to be borne in mind that we are at the same time dealing with a work of art where an idea could often be more important than a form. Conversely, we can scarcely expect the appearance on coins of any fish type with which the local people were not familiar.[47]

Notes

1 For a helpful overview, see Zograf 1951, 56-71; Kraay 1976, 2-5.
2 Cf., however, Tichij 1917, 6; Semenov-Zuser 1947a, 13; 1947b, 239, who instead of a dolphin saw here a pelamys. Equally dubious is the statement of Semenov-Zuser that the dolphin-shaped cast money of Olbia must indicate the wide-scale consumption of fish by the Black Sea population.
3 Karyškovskij 1982, 87-89.
4 See Toporov 1972, 93; Litvinskij 1975, 253-257; Karyškovskij 1982, 92; Raevskij 1977, 46-49, 53, 119-123; 1985, 109, 111-113, 224 note 21. On fish in Scythian art, see now Michel 1995.
5 Karyškovskij 1982, 98 note 154.
6 On these scenes in the Scythian animal style, see, e.g., Fedorov-Davydov 1975, 23-28; Kuz'mina 1976, 68-70; 1987, 3-12.
7 Kraay 1976, 226.
8 Kutajsov 1986, 94-97; 1991, 46-69; 1995, 39-59; Zolotarev 1986, 88-93; Anochin 1988, 133-136; 1989, Nos. 403-406. However, some specimens of this kind were known already to P.O. Buračkov (1881, 234-235; 1884, 99, No. 11) and A.V. Orešnikov (1892, 11-12, No. 12).
9 Koltuchov 1997, 63. He outlines that it was V.A. Kutajsov, who first put forward this hypothesis, although neither the page he refers to, nor the other pages of Kutajsov's article (1991) reveal it. On the Greek influence upon Scythian animal style, see in general Rostovtzeff 1929, 35; Onajko 1976a, 76-86; 1976b, 71-72.
10 For a brief account of such depictions, see Rostovcev 1913, 45-46; Koltuchov 1997; Gavriljuk (in this volume). On the most recent find of the gold fourth-century BC fish-shaped plaques from Tumulus 1 near the village of Filippovka, see Aruz et al. (eds.) 2000, 120-121, Nos. 58-59.
11 S.G. Koltuchov's excavations of 1993. See Koltuchov 1997, 62-63.
12 The Hermitage collection: Inv. Nos. 25935 and 25936; *SNG BM* 706; The Ashmolean Museum, Heberden Coin Room: May bequest 1961.
13 Naster 1959, 142, No. 850, pl. 45.
14 Schönert-Geiss 1970, pls. 1-2, 5-6, 10.
15 Stolba 1991, 80, 83.
16 It is evidenced *inter alia* by the remains of the numerous fish-salting vats of the first centuries AD. See Tichij 1917, 12-18; Surov 1948, 3-47; Belov 1953, 19-22; Kadeev 1962. One of the second-century AD inscriptions of Chersonesos also mentions the fish-market (ὀψόπωλις). See Semenov-Zuser 1947a, 35-44; 1947b, 244-246.
17 Kuprin 1986, 569-611.
18 Kuprin 1986, 576-580.
19 Ščeglov 1978, 26.
20 Belov 1936, 25; Kadeev 1970, 6.
21 Zograph 1977, 283.
22 Shelov 1978, 87 f.
23 Shelov 1978, 88. Cf. Zograph 1977, 283
24 Koehne 1857, 353; Zograph 1977, 177; Shelov 1978, 88.
25 Brabič 1964, 51.
26 Lebedev & Lapin 1954, 205, table 1, 213.
27 Lebedev & Lapin 1954, 208, table 3, and 213, table 10 for the other Bosporan settle-

ments. See also Nikol'skij 1937, 122 (Elizavetovka). A completely different picture is provided by the fish finds from Berezan and Olbia, where sterlet (*Acipenser ruthenus L.*), Russian sturgeon (*Acipenser güldenstädti Brandt*) and beluga (*Huso huso* L.) certainly prevailed in the catches. See Ivanova 1994, 280-81, tables 1-2.

28 Marti 1941b, 95.

29 Vasmer 1923, 67, 73; Abaev 1949, 170, 175, 193.

30 Tozer 1893, 196.

31 Marti 1941b, 97.

32 Jones 1924, 225, note 6.

33 Jones 1924, 225, note 7.

34 Opp. *Hal.* 3.81: γάγγαμα τ' ἠδ' ὑποχαὶ περιηγέες ἠδὲ σαγῆναι; Pollux 2.169.3: καὶ ὁ περὶ αὐτὸν τόπος γάγγαμον, ἐπεὶ νεύρων ἐστὶ πλέγμα, καθάπερ τὸ δικτυῶδες ὃ νῦν καλεῖται γάγγαμον ἢ ὡς πολλοὶ σαγήνη; Ps.-Zonaras 419.27: Γαγγάμη. ἡ σαγήνη, τὸ δίκτυον; Photius, *Lex.* Γ 3.1: Γαγγάμη· δίκτυον. κυρίως δὲ σαγήνη· ἔνθεν καὶ οἱ σαγηνεύοντες γαγγαμουλκοί; Hesych. s.v. γαγγάμη· σαγήνη ἢ δίκτυον ἁλιευτικόν. καὶ σκεῦος γεωργικὸν ὅμοιον τῇ κρεάγρᾳ; Hesych. s.v. γάγγαμον· δίκτυον. (Aisch. *Ag.* 361) καὶ τὸ περὶ τὸν ὀμφαλὸν τῶν ὑποχονδρίων; *Schol. in Aisch.* Ag. 361a.1: γάγγαμον· δίκτυον.

35 For a more detailed discussion, see Bekker-Nielsen 2002b, 217,

36 See, e.g., Bučinskij 1953, 29; Borisov 1956, 540.

37 Schönert-Geiss 1972, 34.

38 See Wright 1990, 32, 35-38. On her cult in general, see also Hörig 1984, 1536-1581; Bilde 1990, 151-187, with literature. On the other hand, one should agree with Schönert-Geiss (1972, 36) in interpreting one of the most common coin types of Byzantion, showing two basket-shaped objects with an altar in between on the reverse, as torches (see already Head 1911, 270; Fıratlı & Robert 1964, 155-156) rather than fish-traps (see, e.g., Franke 1968, 16-17).

39 Rusjaeva 1992, 104; Dubois 1996, 122, No. 74. Dedicatory graffito of the sixth century BC: Ἀθηνόμα[ν]δρος μ' ἀνέθεκεν Αφροδίτηι Συρίηι.

40 Tolstoj 1953, 24, No. 25; Rusjaeva 1992, 104; Dubois 1996, 122, No. 73. Graffito on the fifth-century BC black-glazed kylix: Ἀφ]ροδίτηι Συρίηι Μητρῴ.

41 *IGBul* I, 8bis: [Θεᾷ Σ]υρίᾳ.

42 For additional evidence from the Black Sea, see Alexandrescu Vianu 1997, 15-32.

43 Cf. also the Olbian dedication made by a woman. See note 40.

44 Due to the lack of an image I omit here a single type of Kallatis of the autonomous period mentioned by Mušmov (1912, No. 222).

45 Concerning this coin type of Anchialos, cf., however, Strack 1912, 207: "Ackerbau und Weinbau verbürgen die Münzen, und auch den Fischfang und die Schiffahrt deuten sie klar an".

46 Cf., e.g., mention of the fish-market (ἰχθυοπώλιον) in the Protogenes decree from Olbia (*IOSPE* I^2, 32 B.4).

47 Notwithstanding the great variety of fish in the Black and Azov Seas amounting to more than 130 different species, less than 15% of it seems to be of commercial importance. The number of species suitable for export is even smaller. The data provided by N.E. Maximov regarding catches along the northern shore of the Black Sea from the Danube to the Kerch Straits in the period around 1910 gives an idea as to its quantitative and qualitative composition (See Andrusov & Zernov 1914). With 11,000 people engaged in fishery there the annual take totalled: flat-head mullet – 18,000 specimens, leaping mullet – 300,000 specimens, golden grey

mullet – 13,525,000 specimens, Atlantic mackerel – 73,880,000 specimens, Russian sturgeon, starry sturgeon, and beluga together – c. 424,000 kg, turbot – c. 512,000 kg, Pontic shad – c. 120,000 kg, Mullus barbatus ponticus – c. 208,000 kg, Black Sea anchovy – c. 1,440,000 kg, zostericola – c. 624,000 kg. This covers all the varieties we find listed in, for instance, the Varna Convention of 1959 concerning fishing in the Black Sea (Convention Concerning Fishing in the Black Sea, Varna, 7 July 1959 [http://fletcher.tufts.edu/multi/texts/tre-0230.txt]). The catch of the other species was minor and was not of commercial importance, which might be of some relevance. This is consistent with the osteological materials obtained from the sites excavated in western Crimea. Among the species reported are golden grey, flathead, and leaping mullets, Russian sturgeon, turbot (*Rhombus maeoticus* Pall.) and others. As has been proved by the studies conducted in the 1960s the last 2,000 to 2,500 years do not reveal any significant changes as to varieties, their proportion, areas of fattening, wintering, as well as periods and routes of migration of the main kinds of commercially viable fish (Ščeglov & Burdak 1965; Burdak 1966; Burdak & Ščeglov 1966; Ščeglov 1969; 1978, 26).

The Archaeological Evidence for Fish Processing in the Black Sea Region

Jakob Munk Højte

The archaeological evidence for fish processing in the Black Sea region in the Greek and Roman period is a vast topic covering finds at a large number of sites (Fig. 1) and with interconnections to several other related issues. The literature is extensive, scattered, and inaccessible, if indeed it is available from libraries in Europe.[1] This study, therefore, in no way professes to be an encompassing treatment of all the archaeological evidence for fish processing. This would indeed be far beyond the scope of one paper. Instead I will try to give an overview of the available evidence and present a selection of the most interesting finds and studies. Some of these, like the processing facilities at Tyritake and Myrmekion, are well-known, while others may give a broader perspective on the variety of uses of the fish resources. The aim will be to give an idea of what the archaeological material can reveal about the scale and organisation of the fish processing industry. Throughout I will try to point out some of the shortcomings of the evidence, which I think have not been emphasized enough in the literature, and also point to some areas, where I think scholars have jumped too readily to conclusions.

1. *Types of archaeological evidence*

There exists a wide variety of archaeological evidence that relates to commercial fishing and fish processing. It can be grouped comprehensively in the following manner:

- *Fishing equipment* (net weights, floaters, sinkers, hooks, wrecked fishing vessels, tools for making and repairing nets, and the nets and fish traps themselves)
- *Watchtowers* (σκοπιά)
- *Fish remains* (bones, scales)
- *Processing facilities* (for pickling, salting, smoking and drying. Salt works)
- *Transportation equipment* (shipwrecks, amphorae)
- *Descriptive sources* (epigraphy, coins)
- *Pictorial representations* (sculpture, terracotta, coins etc.)

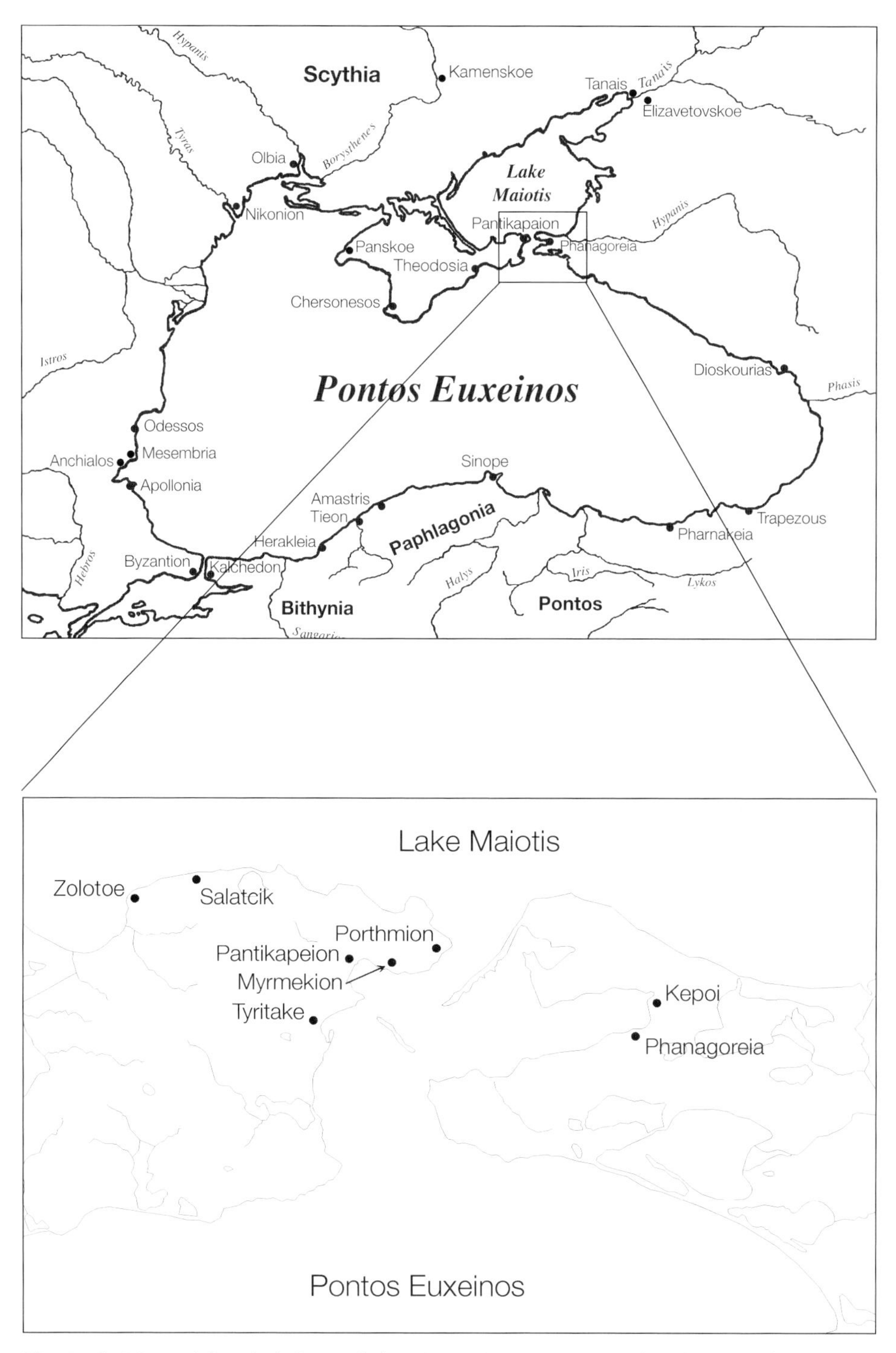

Fig. 1 a-b. Maps of the Black Sea and the Kimmerian Bosporos with indication of the places mentioned in the text.

Not all these types of archaeological evidence will be discussed in the following. Coins with representations of fish have been discussed by Vladimir Stolba in the preceding chapter, and the constributions of Ejstrud, Lund and Gabrielsen below touch upon the vast topic of transport amphorae. As shown by the recent find of a shipwreck off the coast of Bulgaria, to be discussed briefly in the following, this is perhaps one of the most promising fields for advancing our knowledge of the production and distribution of fish products in antiquity. Likewise the sparse – although informative – epigraphic record and the pictorial representations will only be mentioned in passing.[2]

2. *Fishing equipment*

At practically all sites along the northern coast of the Black Sea and around the Sea of Azov, fishing equipment has been reported dating from throughout their entire existence. Particularly frequent are net weights, both lighter ones of clay or lead for throwing nets, heavier ones of regularly shaped stones for dragging nets (Fig. 2),[3] and sinkers of larger stones or even amphora handles used for the same purpose.[4]

Less common are hooks (Fig. 3a),[5] harpoons,[6] and equipment for making and repairing nets: for example bone and bronze needles (Fig. 4).[7] The nets, fish traps, and floaters for keeping the nets afloat have normally not survived, due to poor preservation conditions. An exception to this rule is a small part of a net found in Nikonion.[8] We do, however, have a few sculptural representations of these types of fishing equipment. One example from the Black Sea Region, a terracotta from Kepoi, represents a resting fisherman, with a basket for fishing at his feet (Fig. 3b).[9]

The problem with using fishing equipment to calculate the scale of the activity is of course the need to determine what type and what amount of

Fig. 2. Net weights from Elizavetovka. Left stone weights, right clay weights (after Marčenko, Žitnikov & Kopylov 2000, figs. 75-76).

equipment needs to be present for us to conclude that the fishing carried out did not merely supply a local market for immediate consumption, but was geared to large scale production and export. In many of the storerooms or work areas accompanying identified processing facilities,[10] fishing equipment has been discovered, but it seems virtually impossible to distinguish this equipment from that employed by fishermen catering for a local market for fresh fish. At Porthmion at the entrance of the Kimmerian Bosporos for example, hooks and net weights of the third to the first century BC are found in great numbers,[11] and Gajdukevič takes this as proof of large scale fishing activity.[12] But is this alone sufficient evidence? He draws similar conclusions for the fortified farming and fishing settlements established along the Sea of Azov in the first to third century AD, for example Semjonovka, where hooks, needles, net weights, and fish remains are very common.[13] At Elizavetovka on the Don, fishing equipment is known from as far back as the 4th century BC but for some odd reason no hooks have turned up, although these are known from nearby Tanais in the same period.[14] In this last instance we have reason to believe that processing and export actually took place during this period. This could, however, not have been deduced from the recovered equipment alone.

Fig. 3a. Bronze hook from Panskoye I/U7 in Čornomors'ke Museum (photo: Jacob Munk Højte).

Fig. 3b. Terracotta figure of a resting fisherman found at Kepoi. Now in Taman' Museum (photo: Jacob Munk Højte).

A further factor to be considered is the strategy adopted by the fishers.[15] Ordinarily most fishers in smaller villages may have had fishing as a part time occupation to supply the local market, but in the event of an exceptionally good catch, or in periods of migrating schools of fish, they might have delivered the catch at the nearest salting facility for processing and export. The transition between subsidiary and commercial fishing may therefore have been very subtle, and would not be traceable archaeologically. Nearer to the larger centres, where the demand for fresh fish was greater and where fish processing on a larger scale took place, fishing probably often constituted a full-time occupation, and the investment in equipment was consequently greater. Boats in particular would require a substantial turnover to give a return on the investment. So far no wrecked fishing vessels have been found to compare with the well-preserved boat recovered at Portus which, as evidenced by a built-in well box, clearly fulfilled a demand for fresh fish.[16] It is not entirely impossible, though, that the anaerobic conditions in the Black Sea at depths greater than 200 m will one day reveal an excellently preserved example, but most probably we will not be able to determine the type of fishing for which the vessel was used. Thus, fishing equipment in general can be a good indication of whether fishing took place at all, as for example at the

Fig. 4. Bone needles for repairing nets from Elizavetovka (after Marčenko, Žitnikov & Kopylov 2000, fig. 77).

Scythian sites along the Dnieper (Borysthenes) and Don (Tanais) where, to some degree, fishing seems to be linked to a shift from a nomadic to a semi-settled population.[17] On the other hand, it does not allow us to determine whether the fish actually caught with the equipment was consumed fresh or processed.

A point worth noting is that there does not seem to be any development in fishing technology throughout the period under consideration, judging from the available archaeological evidence. For example pyramidal shaped net weights remain unchanged from the Archaic to the late Roman period and are the most commonly occurring fishing equipment[18] throughout this time. Changes could, however, have taken place without leaving any traces. Firstly, the same equipment could be employed in different ways with a significant effect on efficiency. Multiple hooks could be attached to the same line or net weights could be applied to nets of different sizes. Secondly, there is all the equipment made of organic material that has not been preserved. The degree of use and the size of fishing vessels and the construction of fishing nets are of great importance for the productivity of fishing. In these matters we possess very little knowledge.

3. *Watchtowers*

Watchtowers or lookout posts (σκοπιά) known from literary sources belong to the category of potential evidence, since none have as yet been identified. They evidently served to give advance warning when schools of migratory fish were approaching. A simple shed may have served the purpose, but sometimes they seem to have been of a more permanent nature. Strabon informs us that even in his day the Klazomenians had a watchtower on the sea of Azov.[19] As for the southern shore of the Black Sea we hear about certain places where shoals of fish, particularly tunny, were caught on a regular basis. Strabon mentions Trapezous, Pharnakeia, and Sinope as the main fishing grounds (πηλαμυδεῖον),[20] and Athenaios cites Euthydemos for calling Byzantion "the mother of tunny".[21] At these places it would seem likely that some sort of lookout post existed with a device for signalling the nearest harbour about the approaching schools of fish.[22] The amount of fish caught during migration would clearly exceed the quantity that could be consumed locally in a fresh condition, thus some form of preservation would be required.

4. *Osseous remains and scales*

Osseous remains and scales constitute a very large and interesting group of archaeological evidence. When studied meticulously it offers important information about the ichthyofauna in a given area at a given time. By determining the size and age of the fish it is furthermore possible to obtain valuable data about the intensity of exploitation of the resource and possibly the mode of fishing. As far as the actual quantity of fish caught is concerned, the evidence is much more problematic. One can quite easily calculate the weight of the live fish from an assemblage of fish remains.[23] A rough estimate can be arrived at from the observation that the weight of dry fish bones constitutes approximately 5% of the original weight of the fish,[24] and other much more precise methods based on the number and size of bones and scales have been developed. The problem is, however, to determine the amount of unrecovered osseous material from any given processing site. Bones could be removed for a variety of reasons. First of all they could be collected and disposed of elsewhere – possibly as fertiliser, or they could be removed by animals. Far more importantly, however, the bones could be exported along with processed fish. A good example of this is the wreck recently discovered off Varna in Bulgaria, from which a Sinopean amphora was recovered.[25] This had held large chunks of salted catfish, of which only the bones now remain. Since only one amphora was retrieved from the wreck it should be stressed that we do not know yet whether it is representative of the whole cargo. The mode of production employed here, whether the fish were dried or salted and then transferred to the amphora or salted directly in the amphora, caused the fish, archaeologically speaking, to disappear completely from the processing site. The osseous remains found at the site will to a large extent have derived from fish consumed locally. In other instances certain parts of the fish – head, tail or scales – may have been removed before processing, in which case far more reliable data about volume can be obtained. One such example comes from Tarpachi on the Tarkankhut peninsula.[26] In a stratum probably dating from the third to second century BC, a 1 cm thick layer of grey mullet scales was recovered. Probably several catches were brought here for cleaning and further processing.

The osseous remains found in connection with permanent fish processing facilities raise a number of questions. First of all, the amount of bones in the vicinity in no way accounts for the volume of fish we must assume was being processed in order to render the installations profitable. Again we must assume that, to a large extent, the bones were exported along with the fish. Secondly, there is the question of how the bones that were recovered relate to the production in general. In a number of salting vats a layer of fish bones has been identified at the bottom.[27] This regularly passes as evidence for the type of fish being processed, but in fact it need not be representative of anything but the content of the very last batch before production was discontinued, as we must assume that the vats, at least to some extent, were cleaned between batches. The evidence in all instances points to rather small fish such as anchovy, khamsa, herring, and mullet, but these could feasibly have been characteristic only of the later period of the existence of the processing facilities.

One of the most comprehensive studies of the ichthyofauna in the Black Sea area in antiquity concerns the fish bones of Olbia and Berezan in the

Dnieper (Borysthenes) and Bug (Hypanis) estuary.[28] That these waters were important fishing grounds from early times is hinted at by Herodotos, who praised the sturgeon of the Borysthenes, which he says was salted.[29] It has even been suggested that fishing was indeed one of the principal reasons for settling in this area in the 7th century BC.[30] N.V. Ivanova has examined nearly 6,500 bones from these two locations, 4,867 from Olbia and 1,602 from Berezan. The period under consideration stretches from the seventh century BC to the fourth century AD, with the Hellenistic period giving the largest yield. In all, 19 species of five families were identified, with the evidence from Olbia showing the greatest variety. At Berezan 13 species were represented, all of them present at Olbia also. The most striking fact the data reveals seems to be the clear dominance of very large fish: sturgeon, pike and catfish, while smaller fish are under-represented throughout the period. Carp and roach do occur in some quantity, but their importance is definitely secondary. We seem to be dealing with a very specific preference for fish that live in the slow currents of large rivers: not entirely surprising given the position of the two places. It contradicts, however, the evidence of the locations where fish processing facilities of the Roman period have been identified such as Tyritake and Chersonesos. Here, as seen above, migratory saltwater fish dominate. This trade seems to have bypassed Olbia entirely. Instead the most commonly occurring bones in Olbia and Berezan' were those of catfish. The content of the amphora from the recently found shipwreck off the coast of Bulgaria has been identified as catfish, and since catfish are relatively rare among the osseous remains in other areas of the Black Sea except for sites on the Don,[31] it therefore seems likely that the salted fish in the amphora had its origin here.[32] Next in terms of the quantity of bones are the different types of sturgeon: sterlet, beluga, sevriuga, Russian sturgeon and finally pikeperch, all rather large species.

Another interesting feature revealed by the study is the general decrease in the size of the fish throughout the period from the Archaic to the late Roman period, particularly for catfish, which falls from an average size of 1.59 m in the Classical period to under 1 m in the Roman period. Ivanova attributes this to excessive fishing of certain species, and it seems to indicate that fish were not an inexhaustible resource, at least with regard to the larger fish living in the estuaries of the great rivers. Today these species are largely extinct due to modern industrialized fishing techniques.

Another study by Tsepkin and Sokolov[33] concerns the sizes of the four major species of sturgeon: Beluga (*Huso huso*), Russian sturgeon (*Acipenser güldenstädti* Brandt), sevriuga (*Acipenser stellatus* Pallas), and sterlet (*Acipenser ruthenus* Linnaeus) found in archaeological material from sites in the lower Don region from the fourth century BC to the third century AD. Here it is characteristic that the specimens were extremely large. For example, 16 of the beluga found had lengths in excess of 4 m. In the middle ages the average sizes of sturgeon increases, denoting either that fishers went specifically for the larger specimens or that the intensity of fishing was lower, whereby the fish generally lived to a greater age. Again modern comparison shows that industrial fishing methods have reduced the average size considerably and sturgeon now appear on the endangered species list.

In a sample of osseous material from Pantikapaion, Phanagoreia and settlements in the eastern part of the Sea of Azov published by Lapin and Lebedev, the most common fish was pikeperch closely followed by different types of sturgeon. Unfortunately the data are not directly comparable since they belong to different chronological periods. The rather small 2nd century BC sample from Pantikapeion corresponds well with the finds from Olbia. In the 3rd and 4th century AD in Phanagoreia, carp has taken first place at the expense of larger species.[34] However, the samples are too small and from too few contexts for us to determine whether they are coincidental or whether they represent a general tendency towards catching smaller types of fish in the Roman period. Such a shift could very well have been caused by a shift in production methods from salting chunks of large species, as in the shipwreck near Varna, to production of other salted fish products which could be made from smaller fish.

5. *Fish processing facilities*

Lastly we turn to the most prominent of the archaeological evidence, namely the remains of the processing facilities for salted fish products.[35] These consist normally of a series of vats built up or hollowed into the rock, lined with walls and finally waterproofed with *opus signinum* containing a high content of crushed ceramic material giving them a reddish colour. Storage and work facilities are usually found in connection with these vats also. Before introducing the five locations with such salting vats, it is worth considering what, to my knowledge, is the only identified pre-Roman installation for preserving fish in the Black Sea area. It was not intended for salting fish, but instead for smoke-curing fish.

5.1 *Elizavetovka*

The Elizavetovka Settlement southeast of Tanais has been excavated by Russian archaeologists since the 1940s but has only recently undergone proper publication.[36] The excavation shows that fishing played an important role in the economy of the settlement. In some areas of the site large plots were covered with up to 20 cm thick layers of compressed fish bones, and in the periphery of the settlement refuse pits filled with scales and bones have been uncovered.[37] As mentioned above, fishing equipment is found in abundance. Characteristically, fishing in the area only seems to have begun with the establishment of the settlement. During the nomadic or semi-nomadic periods of Scythian culture fishing played a minor role, although there is no doubt that it was practised, cf. Gavriljuk's contribution to this volume. At Kamenka on the

Dnieper, for example, there are traces of fishing activity during the Scythian period.[38] In the fifth century BC, few remains of fish are found at Elizavetovka Settlement, nothing to denote export. In the fourth and third century BC with the growing Hellenization, however, there is a fishing boom. The excavators believe that the amount of fish caught as early as in the first half of the fourth century BC already exceeded local consumption, and from that point onwards, fish must have been one of the foremost export goods. The excavators stress that the fish remains do not primarily derive from refuse deposits in households but rather from semi-industrial (*handwerklich*) production.[39] In the settlement, 36% of all osseous material derives from fish, most commonly sturgeon and carp, but also small amounts of perch and catfish. As seen above, the specimens are quite large, with a catfish – measuring about 2.40 m – as the largest.[40] No tanks for salting fish have been found either at Elizavetovka Settlement or in nearby Tanais,[41] but instead the excavators have uncovered what may have been a smoke-curing installation. It was situated in the northern section of the settlement in an area with a large amount of fish bones. It consists of two chronologically consecutive pits about 1.3 m in diameter with heavily burned sides and bottoms. In the younger, a pile of charcoal was found along with fish bones. What the installation once looked like and what its capacity for preserving fish may have been is impossible to determine. This unique example should remind us that fish preservation on a large scale could take place without leaving significant archaeological traces. This is especially true of the most basic method of preserving fish, namely by drying, since the fish screens made of wood would not survive at all. Thus the amount of archaeological evidence does not necessarily reflect the level of production but rather the prevalent production method.

5.2 *Tyritake*

The most thoroughly studied fish processing installations are those at Tyritake 11 km south of Pantikapaion, excavated by Gajdukevič from the 1930s to the 1950s.[42] A total of 57 salting vats were uncovered in the southern and eastern part of the city. Surprisingly, all the installations lay within the city wall (Fig. 5). The vats are of rectangular shape and partly hewn out of the rock. Typically the sizes range between 2.00×1.40 and 2.50×1.50. Inside and above the rock surface they are built up and covered inside with waterproof mortar (*opus signinum*). Depths range between 1.50 and 2.00 m with a few up to a depth of 3 m. The smallest vat has a capacity of only 3 m^3, while the largest, an irregularly shaped vat in unit B, measures approximately 22.12 m^3. The vats are all grouped in small production units. Three to six vats seem to be the common size. Typically the vats are in a single row or in two rows of two or three. The largest processing complex in Tyritake, situated by itself in the area just inside the southern wall, had 16 vats, four by four, of regular size (3.20×1.70×1.80) giving a total capacity of more than 155 m^3 (Fig. 6). Found

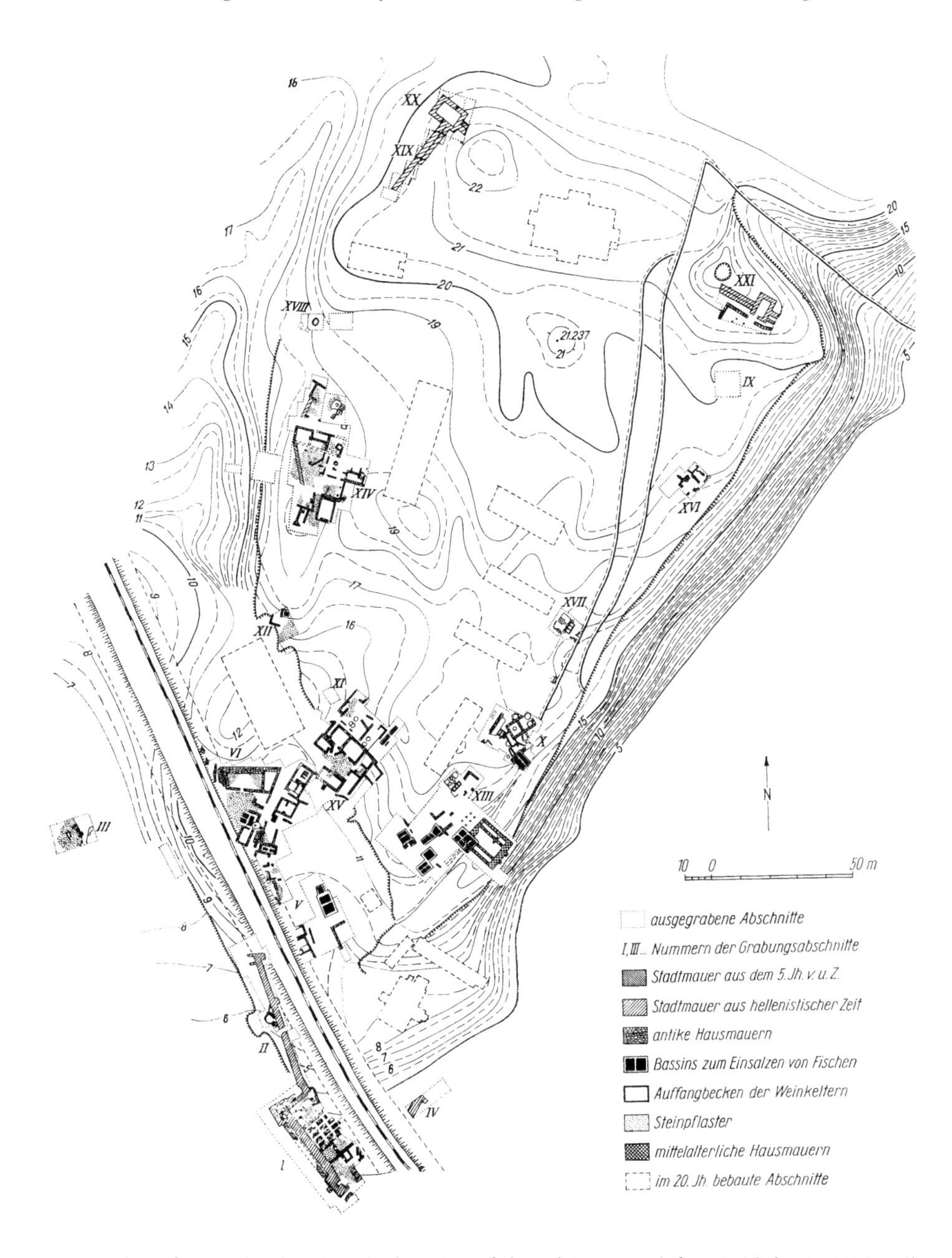

Fig. 5. Plan of Tyritake showing the location of the salting vats (after Gajdukevič 1952, 16).

Fig. 6. The largest salting installation with originally 16 vats located just inside the southern wall. The vats had a capacity of 155 m³ (courtesy of the Photo Archives of IIMK RAN).

at the bottom of these vats were remains of herring. Vats have turned up in several of the excavated sectors, but it is particularly in sector XIII in the eastern part of the city that a high concentration was observed. Here no less than six individual installations were situated, and fish processing seems to have been the only activity in this sector during the first three centuries of our era (Fig. 7-8).[43] The lack of overall planning in the layout indicates that each unit was run separately. Gajdukevič has suggested that all the processing facilities were owned by the Bosporan king and operated by slave labour,[44] but the only evidence to substantiate this claim is the lack of luxurious houses in Tyritake, which in itself does not indicate slave occupants. In the vicinity of the installations, storerooms with *pithoi* are regularly found, and here net weights, fish bones and amphorae abound. To judge from the many tiles found near the salting vats these seem to have been covered by a tiled roof for protection from the weather. According to Gajdukevič's analysis of the finds, all the vats were constructed in the first century AD,[45] but they could possibly have replaced previous processing installations of some sort.[46] Production continues after the third century AD, but the number of vats is reduced and at some point in the fifth century a basilica obliterated at least one of the installations.[47] Whether the rest continued to function remains unknown.

That fish processing was not merely restricted to larger specialized installations is shown by a house of the 3rd to 4th century AD uncovered in sector XV (Fig. 9).[48] Room 1 contained a large *pithoi* with wheat. Other finds include

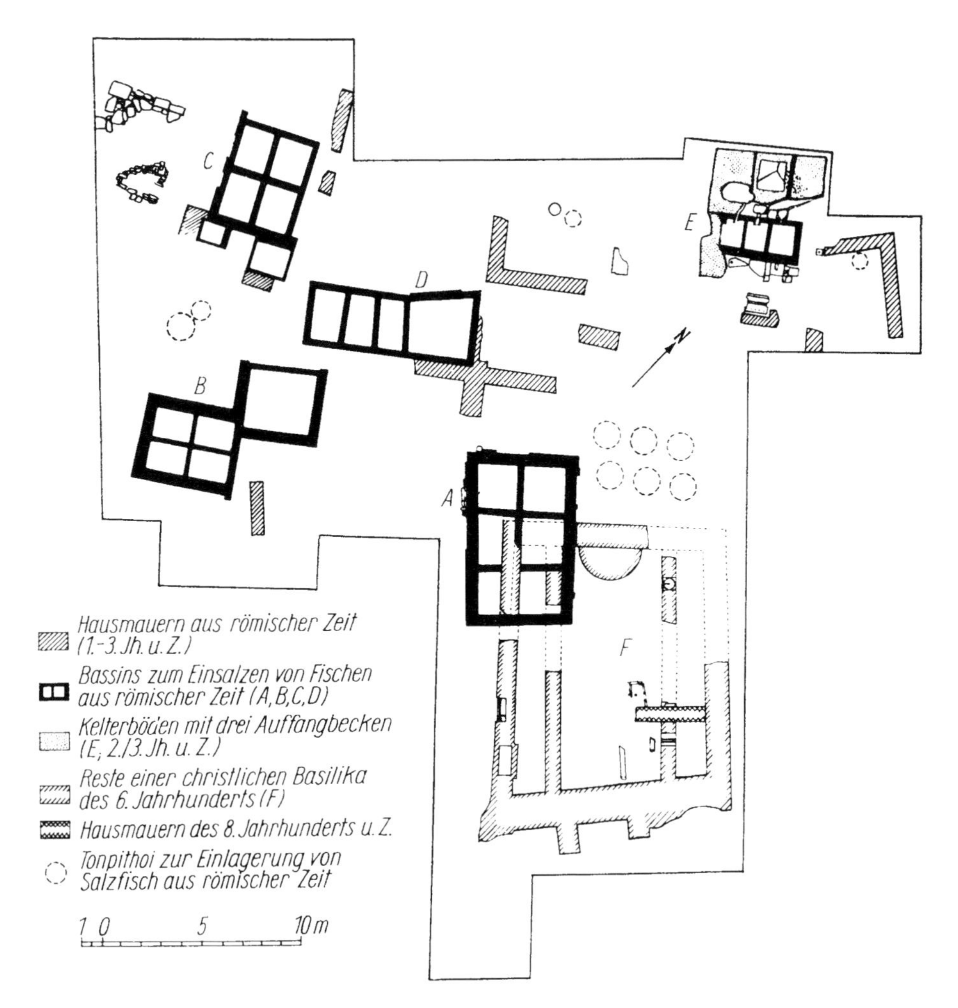

Fig. 7. Plan of sector XIII in Tyritake where a high concentration of salting vats was found (after Gajdukevič 1971, fig. 97).

amphorae, oil lamps and sturgeon scales. Rooms 2 and 3 contained mills and Room 4 seems to have been a storeroom with amphorae. Room 5 may have been a women's room to judge from the spindle whorls and an ivory *pyxis* with red dye. What is interesting is that in almost every room net weights and bone needles were found. Furthermore, outside to the southeast a vat had been built against the wall, which had been used for salting or otherwise processing fish. The house seems to have belonged to a fisherman and his family, who also processed the fish on a very small scale in the household.

The total capacity of the known installations in Tyritake has been calculated to 457 m^3 and they could process up to 365 metric tons of fish simultaneously.[49]

Fig. 8 a-c.
a) Salting unit B in sector XIII in Tyritake.
b) Salting unit D in Sector XIII.
c) Net weights found in the vicinity of salting vats in Sector XIII (courtesy of the Photo Archives of IIMK RAN).

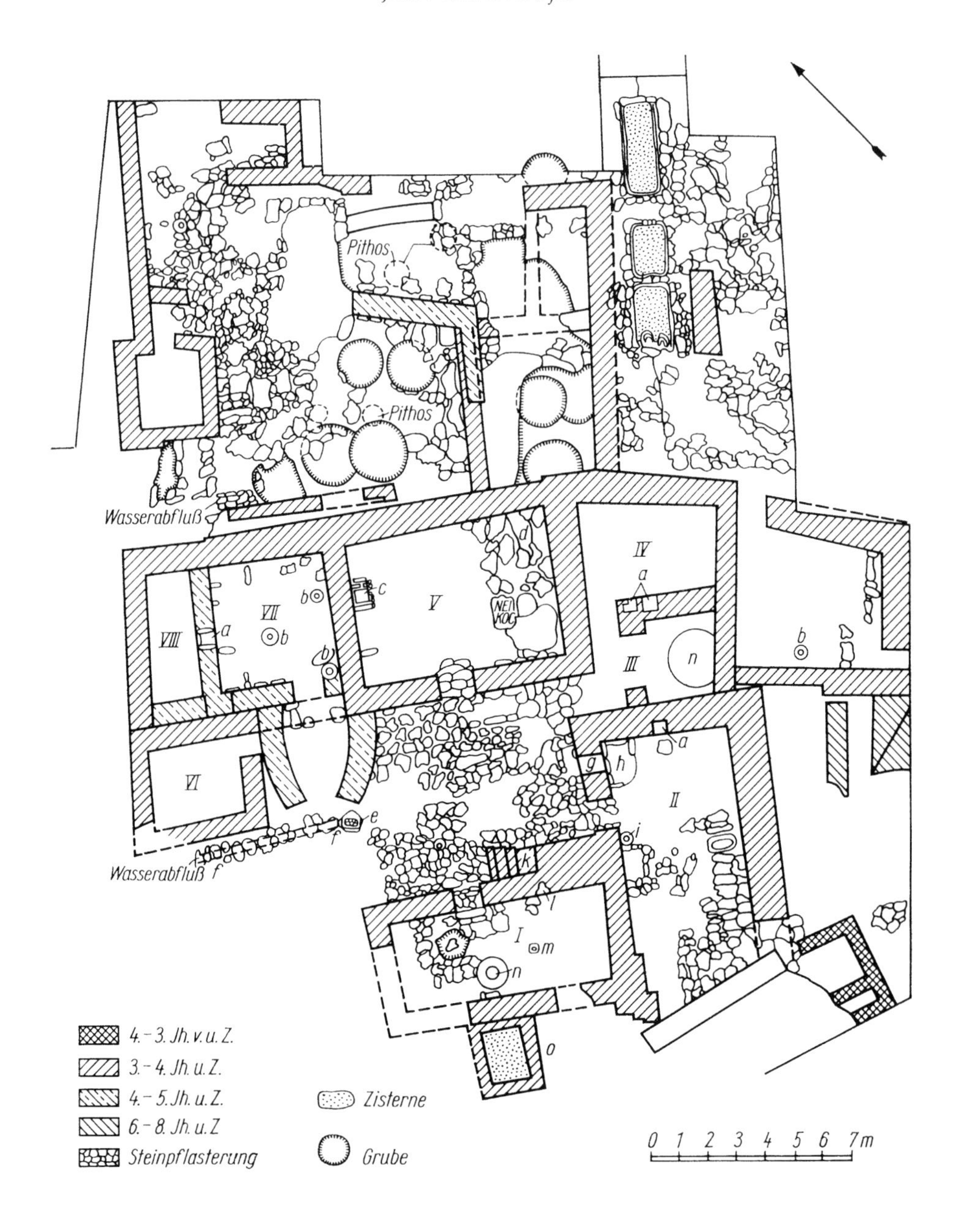

Fig. 9. Plan of house of the 3rd-4th century AD with small salting vat built onto the south-western wall (after Gajdukevič 1971, fig. 108).

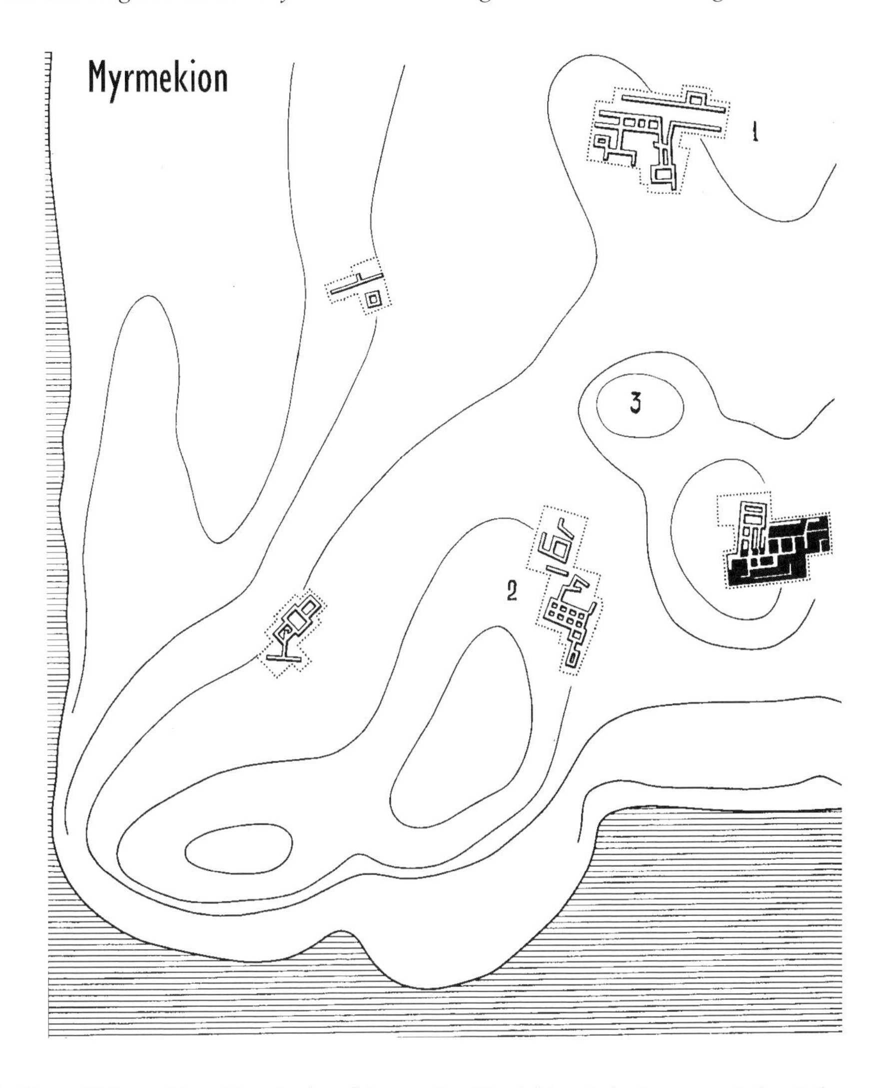

Fig. 10. Plan of Myrmekion. The single salting unit with eight vats in two rows is located in area 2 (after Gajdukevič 1952, 136).

5.3 *Myrmekion*

In Myrmekion a short distance to the east of Pantikapaion another single fish processing installation dating to the 2nd-3rd century AD was excavated by Gajdukevič (Fig. 10).[50] It consists of eight vats in two rows of four, each 3.00×2.70×1.80 m with a total capacity of about 116 m^3, accompanied by a storage room with a number of large *pithoi* (Fig. 11). To judge from the thick layer of bones at the bottom of some of the vats the last catch, at least, was anchovy. Other bones in the area include those of sturgeon. The construction of the vats is similar to those at Tyritake, but finds in the vicinity help to

Fig. 11. Salting unit in Myrmekion with a capacity of 116 m^3 (courtesy of the Photo Archives of IIMK RAN).

shed further light on the production process. The large flat limestone slabs recovered may have been used to press down the fish into the salt solution (Fig. 12). A slightly conical ceramic vessel interpreted as a sieve for extracting the fish from the brine in the vats, or possibly used in the production of *garum*, was also recovered in one of the *pithoi* (Fig. 13).[51] To my mind, however, the holes in the side of the vessel are just ordinary repair holes rather than holes for drainage or for attaching a rope. What purpose this unusual vessel served remains unclear. Only a relatively small area of the town has been excavated, so it is quite possible that further excavation would reveal more installations.

5.4 Chersonesos

The city with the largest known capacity for fish processing was Chersonesos.[52] The installations have not, however, received quite the same thorough attention as those in Tyritake. An exception is a house in block XV-XVI in the northern central part of the town where a Hellenistic house in the first century AD was turned into a small fish processing facility.[53] According to Kadeev there are about 90 salting vats of all periods, predominantly in the harbour area, with a total volume of some 2000 cubic metres.[54] A recent publication

Fig. 12. Finds from the vicinity of the vats in Myrmekion. Note the tiles that may have belonged to a protective roof, and the limestone blocks that were used to press down the fish during processing (courtesy of the Photo Archives of IIMK RAN).

Fig. 13. Ceramic vessel thought to have functioned as a sieve (courtesy of the Photo Archives of IIMK RAN).

raises the figure to more than 100 vats.[55] To judge from the osseous remains found in the salting vats, anchovy (khamsa) seems to have been the primary catch.[56] Herring have also been identified.[57] The facilities in Chersonesos show a number of peculiarities. First of all they tend not be organized in larger units, but rather appear solitary or in groups of two or three at the most in what seem to be private houses. The individual vats also tend to be larger than those of Tyritake, particularly as regards their depth: 3 m or more does not seem to be unusual. Many of the vats are hewn out of the rock, lined with stones and finally waterproofed with *opus signinum* as at Tyritake (Fig. 14). In contrast to the situation in Tyritake, there is no evidence to show that the vats were covered by roofing. One scholar mentions pear-shaped vats, but there is every reason to doubt that these had the same function. This shape would have been quite impractical for the purpose. Rather they were probably water cisterns.[58] Beside the vats there are nearly always storerooms containing several *pithoi* (Fig. 15). The ceramic evidence points to a construction date in the first to second century AD for most of the installations, and production probably continued throughout antiquity.

Kadeev has calculated the yearly capacity to at least 3000 to 3500 metric tons,[59] but compared to the estimates for Tyritake and Myrmekion this figure is very conservative.

Fig. 14. Cistern A in a house in block XV-XVI in Chersonesos from the first to second century AD (after Belov & Strželeckij 1953, 80, fig. 44).

Fig. 15. Pithoi in a storeroom in the house in block XV-XVI in Chersonesos (after Belov & Strželeckij 1953, 79, fig. 43).

5.5 Zolotoe and Salatčik

More recently two further fish processing installations at Zolotoe and Salatčik on the Maeotis side of the Kerch peninsula have been identified.[60] These show many of the characteristics of the installations in Tyritake with regard to organisation and construction. The better preserved installation at Zolotoe probably consisted of four large tanks, two of which are fully preserved, the other two having been partially washed away by the sea (Fig. 16). The larger vat measures about 23.5 m^3 making it the largest known example in the Kimmerian Bosporos. If we assume that the two vats which did not survive were of similar size, the complex had a capacity of over 83 m^3. According to the estimates given for production in Tyritake, the complex could process approximately 65 metric tons of raw fish per filling. Vinokurov proposes eight annual productions and arrives at a capacity of 530 metric tons of fish. For this process 125 tons of salt would have been required.[61] Operating this facility (catching the fish, acquiring salt, filling vats, loading amphorae etc.) would have required considerable labour and would certainly have contributed significantly to the economy of this small community.[62] As at Tyritake, a storeroom was connected to the complex. Found within were pits for *pithoi*, one of which was still in situ. This *pithos* with a capacity of about 1000 litres contained fragments of herring bones. Found on the floor and in the pits were shells, fish bones and scales (unfortunately not specified), and fishing equipment – including net sinkers made from amphora handles. The amphorae, which constitute over 90% of the diagnostic profiles, and the *sigillata* found in the complex, date from the second and third centuries AD. This means activity here started somewhat later than at the installations in the larger cities of Tyritake, Myrmekion and Chersonesos.

The processing unit in Salatčik is far less well-preserved (Fig. 17). It consists of at least two – seemingly – very large tanks, but neither their size nor depth can be determined precisely, as they have been almost obliterated by houses of the fourth century AD.

The importance of these two new processing installations consists primarily in the fact that they show that fish processing within the Bosporan Kingdom may have been far more dispersed than the previous finds at Tyritake and Myrmekion would suggest.[63]

One last aspect of the preservation of fish needs to be mentioned, namely the amount of salt required for salting fish. Vinokurov, as mentioned above, suggests that the annual amount of salt needed at Zolotoe was over 125 metric tons. The amount of salt needed in Chersonesos, Tyritake, and Myrmekion would have been far greater. Kadeev has calculated that the salting vats in Chersonesos required around 800 metric tons and possibly more during peak years.[64] So far, very little work has been done concerning salt production in the Black Sea region. There are plenty of references to salt extraction taking place around the Black Sea. Herodotos (4.53) and

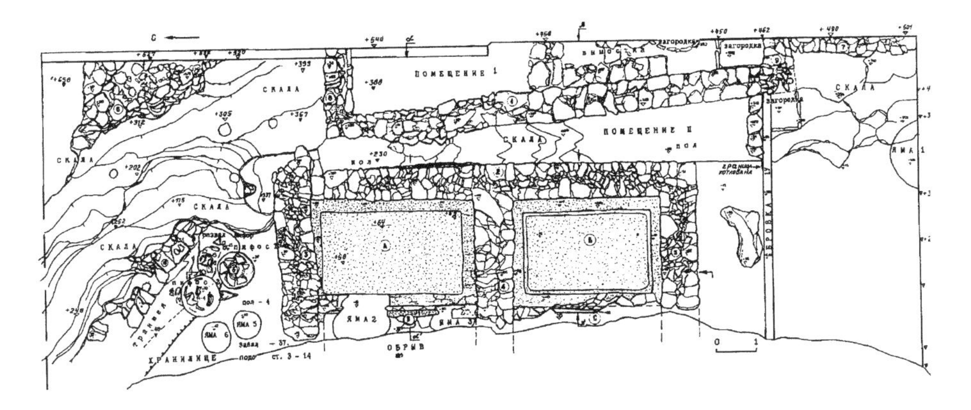

Fig. 16. Plan of the salting installation at Zolotoe (after Vinokurov 1994, 158-159, fig. 2).

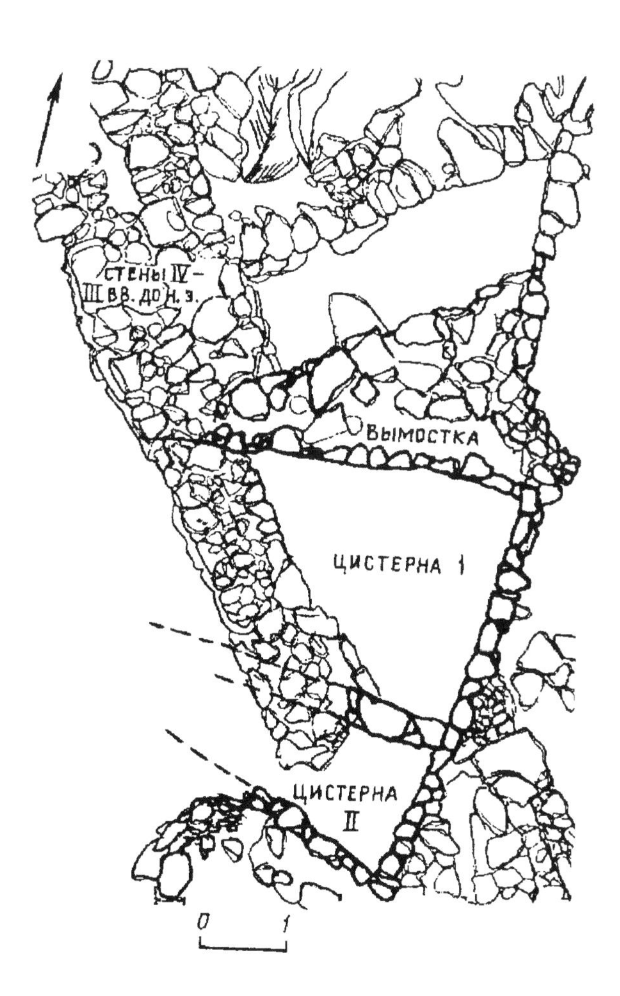

Fig. 17. Plan of the salting installation at Salatčik. (after Vinokurov 1994, 168, fig. 10.2).

Dion Chrysostomos (36.3) mention salt beds near Olbia in the Bug-Dnieper Estuary, which apparently also served the Crimea. Strabon (7.4.7) refers to salt works south of Chersonesos, and according to him salt was bought by local tribes in Dioskourias (11.5.6). He further relates (12.3.12; 12.3.39) that the river Halys took its name from the salt mines it flowed past in Ximene. To my knowledge none of these production sites have ever been identified. Only around Chersonesos have possible sites for salt extraction in antiquity been identified, primarily on the basis of the existence of later activities.[65] The changing landscape may of course have eradicated most of the evidence, but some traces of the infrastructure such as earthen dams, roads, and possibly jetties are likely to have survived. Near Pomorie in Bulgaria, salt is still being produced and the history of the saltworks can be traced back to at least medieval times. Whether Apollonia and Mesembria exploited the salt beds commercially in antiquity remains unclear.[66]

6. *Conclusion*

Practically everything discussed above concerns the northern part of the Black Sea. What about the other areas of the Black Sea? Here the situation is altogether more disappointing. We have plenty of literary and epigraphic evidence for commercial fishing and processing all around the Black Sea, but hardly any archaeological evidence to match it.[67] The western coast is fairly well explored, but no processing facilities have yet been identified. Along the southern coast of the Black Sea, disappointingly few excavations and surveys have been carried out. One of the most promising sites is at Sinope, where an intensive survey has been carried out over the past years. However, the only evidence for fish processing is a single tank near the port of Armene, which could easily be later than the Roman period.[68] Further research in Northern Turkey may change this situation significantly. Thus, necessarily, the conclusions drawn from the presented material only concern the northern coast.

As shown, plenty of evidence exists for fishing in the form of fishing equipment throughout the period under consideration, both at the Greek and Scythian settlements. The early literary sources also repeatedly describe the different fish being caught and fish products being turned out at different places around the Black Sea. Before the Roman period, however, we have very little archaeological evidence for fish processing. This can be explained in several ways. Firstly, it may not have taken place at all: the fish caught were all consumed fresh. Secondly, the Roman salting installations could have obliterated earlier Greek ones at the same locations. Thirdly, the production methods employed during the Greek period simply left very few traces. The first option hardly seems credible in the light of the literary evidence. Just to mention a few examples, Demosthenes refers to salted fish in transit from Pantikapaion to Theodosia (35.34; cf. Gabrielsen and Lund, in this volume), Polybios speaks of salted fish being exported to Rome (31.25.5), and Strabon seems to think that the export of salted fish from the Kimmerian Bosporos during the Roman period was merely a continuation of an old practice (7.4.6). Neither is the second hypothesis adequate, as vats of an earlier period are unlikely to have been completely obliterated by later buildings. The second century AD installation at Salatčik, for example, was still clearly discernible under the fourth century AD house. The last option seems the most feasible. Drying, smoke curing or salting, for that matter, could be practised on a large scale without leaving significant traces. In this context the smoke-curing pit at Elizavetovka Settlement is very important, as it documents this practice for the first time. The scantiness of the evidence makes it next to impossible to estimate the volume of the production, but in such situations it is very easy to adopt a much too pessimistic view. Further research into shipwrecks may show how common the transport of fish products was before the Roman period compared to other commodities.

The emergence of the salting installations in the first century AD in the Bosporan Kingdom and in Chersonesos certainly signifies an important change. But the question remains whether it was only a change of production mode or whether it actually changed the quantity of processed fish. I would be sceptical of Gajdukevič's interpretation, that in the Greek period the export of fish was restricted to finer fish as luxury commodities, while in the Roman period cheaper pickled fish such as herring and anchovies were exported to meet the demands of a broader consumer market, and perhaps to the Roman army stationed on the Danube and in Asia Minor.[69] To my mind the one need not exclude the other.

During the Roman period, fish processing seems to have been big business, but can we estimate the volume of the production more precisely? For the installation at Zolotoe, Vinokurov estimated that 560 metric tons of raw fish could be processed annually, and suggestions for the total production from the northern Black Sea area exceed 20,000 tons. Such calculations are fraught with uncertainties. First of all we do not know how large a percentage of the salting installations once in existence have actually been found. As for the individual installation we can calculate the maximum capacity of each vat, but we do not know whether they were always filled to the top. Neither does our limited knowledge of the actual process leave us any clues as to the length of the process. Furthermore we have no reliable estimates for the availability of fish throughout the year. Production could have been seasonal. The suggestion of eight fillings a year is therefore nothing but an educated guess. Despite these reservations, I think it can still be concluded that fish processing accounted for a significant portion of the economy of Chersonesos and the Bosporan Kingdom. The prominence of the salting vats in the urban architecture, particularly in Tyritake, testifies to the importance of this trade.

Notes

1 I wish to thank Vladimir Stolba for helping me achieve the correct interpretation of a number of articles in Russian.
2 For representations of fish in Scythian art, see Gavriljuk (in this volume).
3 For the net weights at Elizavetovka, see Marčenko, Žitnikov & Kopylov 2000, figs. 75-76. Of particular interest among the many publications that include net weights is a study of the net weights of Phanagoreia from the 6th century BC to the 4th century AD carried out by Onajko (1956, 154-163).
4 Vinokurov 1994, 163.
5 Tichij (1917, figs. 4-5) and Carter (2003, 86) show a selection of hooks from Chersonesos. Čornomors'ke Museum displays a few fishing hooks from Panskoye I, building complex U7.
6 A small harpoon for fishing from Panskoe I/U7 can be seen in the museum in Čornomors'ke.
7 Marčenko, Žitnikov & Kopylov (2000, Fig. 77) has a fine selection of bone needles. Mack (2003, 86) shows bronze needles in Chersonesos Museum. Kruglikova 1963, 43-51.
8 Brujako 1999, 52-53 & fig. 17.
9 Sokol'skij 1968, figs 5 & 5a.
10 Gajdukevič 1952, 59; Vinokurov 1994, 154-170.
11 Kastanajan 1959, 203-207.
12 Gajdukevič 1971, 184.
13 Gajdukevič 1971, 201, 411-412. For Semjonovka, see Kruglikova 1963, 43-51.
14 Marčenko, Žitnikov & Kopylov 2000, 179. For Tanais, see Šelov 1970, 186.
15 Bekker-Nielsen 2002a.
16 Testaguzzi 1970, 143-44.
17 Marčenko, Žitnikov & Kopylov 2000, 175-176 examples with references. See also Gavriljuk in this volume.
18 Onajko 1956, 154-163.
19 Strabon 11.2.4. For a thorough survey of the literary sources for fishing and fish processing in the Black Sea region, see Curtis 1991, 118-129.
20 Trapezous: Strabon 7.6.2; Pharnakeia: Strabon 12.3.19; Sinope: Strabon 7.6.2, 12.3.11.
21 Athenaios 3.116b. See also Dumont 1976-1977, 96-117.
22 For similar installations in Italy at Cosa and Populonia that certainly functioned as lookout posts (*thynnoskopeia*), see Strabon 5.2.8 and 5.2.6.
23 Casteel (1976, 93-122) investigates four methods of estimating fish size and weight from bones.
24 Casteel 1976, 119-122.
25 Reported with photographic documentation at the National Geographic home page at: http://news.nationalgeographic.com/news/2003/01/0110_030113_blacksea.html
26 Ščeglov 1969, 128-130.
27 Chersonesos (anchovies, khamsa): Belov & Strželeckij 1953, 80; Kadeev 1970, 14. Tyritake (herring, mullet): Gajdukevič 1952, 59. Myrmekion (khamsa): Gajdukevič 1952, 207.
28 Ivanova 1994, 278-283.
29 Herodotos 4.53.

30 Domanskij & Marčenko 2003, 29.
31 Tserkin & Sokolov 1971; Casteel 1976, 130-132; Marčenko, Žitnikov & Kopylov 2000, 179. According to Lapin and Lebedev (1954, 197-214) catfish occurs at Pantikapaion but is very rare at Phanagoreia and at sites in the eastern part of the Sea of Azov.
32 Andrei Opait has kindly informed me that the amphora in question is a type Zeest 85 similis of the 2nd to 3rd century AD and not as stated in *National Geographic*, a Sinopean amphora of the 5th to 4th century BC.
33 Tserkin & Sokolov 1971; Casteel 1976, 130-132.
34 Lapin & Lebedev 1954, 197-214.
35 Curtis (1991, 6-26) discusses the terminology for the range of salted fish products and their method of production. For a very early study of fish processing in the northern Black Sea area, see Köhler 1832. Also of note are Minns 1913, 440 and Danov 1962.
36 Marčenko, Žitnikov & Kopylov 2000, 175-181 concerning fishing and fish processing.
37 Marčenko, Žitnikov & Kopylov 2000, 175.
38 Grakov 1954, 144.
39 Marčenko, Žitnikov & Kopylov 2000, 177.
40 Marčenko, Žitnikov & Kopylov 2000, 179.
41 Šelov (1970, 186) has proposed that drying fish was the primary method of preservation at Tanais.
42 Gajdukevič 1952a, 15-134; Gajdukevič 1971, 376-378.
43 Gajdukevič 1952a, 55-72.
44 Gajdukevič 1971, 185-186.
45 Gajdukevič 1952a, 15-134.
46 Curtis 1991, 126.
47 Gajdukevič 1971, 485.
48 Gajdukevič 1971, 408.
49 Marti 1941c, 103. Marti 1941b, 94.
50 Gajdukevič 1952b, 135-220; Gajdukevič 1971, 378.
51 Gajdukevič 1952b, 207, fig. 125 & 126.
52 Tichij 1917; Semenov-Zuser 1946, 237-246; Mongait 1959, 199; Brašinskij 1968, 96-97; Kadeev 1970, 5-26. Kadeev & Ryzov 1973, 76-80; Romančuk 1973, 45-53; Romančuk 1977, 18-20; Mack 2003, 86. Undoubtedly not all tanks functioned simultaneously. According to Romančuk (1977, 18-20) many of the tanks only contained material from the medieval period.
53 Belov & Strželeckij 1953, 32-236.
54 Kadeev 1970, 14.
55 Mack 2003, 86.
56 Belov & Strželeckij 1953, 59-60.
57 Kadeev 1970, 14.
58 Mongait (1959, 188) is probably referring to the cisterns in block XV-XVI published by Belov & Strželeckij (1953, 32-236). For their function, see Mack 2003, 86.
59 Kadeev 1970, 15.
60 Vinokurov 1994, 154-170. Zolotoye excavated in 1987 and Salatčik in 1990.
61 Vinokurov 1994, 166-167.
62 For estimates, see Vinokurov 1994, 167.
63 After completion of this manuscript, I was informed by staff in the museum in Anapa, ancient Gorgippia, that cisterns possibly for salting fish were found in

excavations during the construction of the modern harbour, but that no record of the excavations exist. It may be these cisterns that Alekseeva refers to without reference in her book: *Anticnyj gorod Gorgippija* (Moscow 1997) 168.

64 Kadeev 1970, 25. It is however, as shown above, not entirely clear exactly how many vats were operating simultaneously in Chersonesos.

65 Kadeev 1970, 20-26.

66 Hoddinott 1973, 221, but without references. *RE* 1, s.v. Anchiale 1, col. 2103 does not mention salt production. Today a salt museum is being constructed at the site.

67 Curtis (1991, 118-129) offers a general survey of the sources. Trapezous, Sinope, Amastris, Tieon, Herakleia Pontike, Kalchedon and Byzantion are among the cities noted for their fish products, and an association of fishermen is known from Odessos (*AE* 1928, 146).

68 I wish to thank Alex Gantos, Assistant Director of the Sinop Regional Survey, for sharing this information.

69 Gajdukevič 1971, 376.

A Fishy Business: Transport Amphorae of the Black Sea Region as a Source for the Trade in Fish and Fish Products in the Classical and Hellenistic Periods

John Lund and Vincent Gabrielsen

> Salted, dried, and pickled fish, the staple food of the Greeks, was imported in large quantities into Greece, Egypt, and probably Syria from the Pontic regions and from Sicily
> (M.I. Rostovtzeff 1941)[1]

> This ill-conceived picture of the Pontic fisheries has, unfortunately, found a wide audience
> (T.W. Gallant 1985)[2]

The notion that fish and fish products could have been carried in transport amphorae produced in the Propontis, i.e. the Sea of Marmara, or along the northern coast of Asia Minor may be traced back to articles published by J.L. Stoddart in 1850 and 1853. These articles are, incidentally, among the first to treat transport amphorae as evidence of trade.

In 1850, Stoddart wrote:

> That the ancient commerce of Alexandria should have connected itself with the towns on the Propontis and its dependant straits, is very intelligible. Wine, which was plentiful and good on the Asiatic shore, was, no doubt, sought there; but the riches and consequence of all those places depended mainly on their fisheries … They were to the Greeks what Newfoundland is to us. Nowhere else was fish more abundant, nowhere so excellent … The smaller kinds entire, and the larger kinds were carved into lumps, with salt strewed between the layers, they were packed in amphorae, … or in larger vessels (πίθοι), and so exported.[3]

In 1853, the same scholar suggested that salted fish "was the leading commodity which the ancient maritime towns on the south side of the Euxine exported in *diotae* to those on the north side, where its superiority to their coarser and less sapid sturgeon must have procured for it an extensive demand".[4]

Stoddart did not present any positive arguments to support his hypothesis, and he did not specify which centres he imagined to be involved in this traffic. Still, his theory is highly relevant to the theme of this workshop, and the aim of this paper is to present and discuss the archaeological evidence for the use of transport amphorae of the Black Sea region as a possible source for the trade in fish in the Classical and Hellenistic periods.[5]

1. *Prolegomena*

Before turning to the Black Sea region, it may be mentioned that transport amphorae have been the subject of intensive research in the last decades,[6] no doubt because such vessels were "above all ... containers used in seaborne commerce", which "provide us ... with direct witness of the movement of certain foodstuffs which were of considerable economic importance".[7] David Peacock and David Williams wrote these words in their study on "Amphorae and the Roman economy", which was published in 1985. Since then, our knowledge about the typology and chronology of transport amphorae has advanced significantly, and research in amphora kilns and workshop facilities has also made great strides forward. Thanks to new publications of quantified contextual evidence, we may now map the regional and interregional distribution of many amphora types with some confidence.

In other respects, however, progress has been less marked. This is, for instance, the case with regard to determining the contents of the amphorae – especially those predating the Roman era.[8] This question is surely of the utmost importance, if we want to use amphorae as a source for ancient trade and economics. Several sources of information about this matter are at our disposal: 1) residue analyses by means of gas chromatography-mass spectrometry, 2) finds made inside sealed amphorae found in shipwrecks, 3) indications from *graffiti* or *dipinti*, which may, however, be secondary and hence misleading,[9] 4) iconographic evidence from amphora stamps and representations on coins and other media, and 5) ancient written sources. On the basis of all of this, a consensus of sorts emerged, which was formulated by Carolyn Koehler in 1996: "wine has been nominated as the chief export in amphoras from a number of Greek cities, including Chios, Kerkyra, Knidos, Kos, Lesbos, Mende, Paros, Rhodes, Sinope (and other sites in the Black Sea) and Thasos".[10]

In recent years, however, a more subtle approach has emerged: Yvon Garlan has questioned whether each individual amphora type did, indeed, only carry one type of commodity, as was hitherto commonly assumed.[11] Also, Mark Lawall generally refrains from speculating about the content of the amphorae in his illuminating analyses of amphora finds at Ilion,[12] even if he associates the Chian amphorae found in Athens with trade in Chian wine.[13]

2. *Amphorae produced in the Black Sea Region*

The main centres involved in the production of transport amphorae in the Black Sea region in the Classical and Hellenistic period were Herakleia, Amastris, Sinope, Dioskourias and Chersonesos.[14] The amphorae produced in these centres have been well studied by several generations of Russian and other scholars, who have mainly concentrated on elucidating their typology, chronology and stamps.[15] The question of their contents has apparently not been at the forefront of research.

Still, it has been claimed that amphorae from the Chersonesos contained "cheap local wine",[16] and perhaps also grain,[17] and that those made at Amastris carried "olive oil and salted olives".[18] The site of Herakleia Pontike has been characterized as "one of the greatest wine exporters to the North Black Sea region" and it is assumed that amphorae from there contained wine.[19] As for the amphorae from Sinope, Nicolae Conovici expressed the opinion that they mainly contained wine,[20] whereas Vladimir Kac and others contend that they "most probably contained olive oil rather than wine; however, one cannot rule out wine".[21] Ancient literary sources refer to the cultivation of both grapes and olive trees at Sinope.[22]

The attributes seen on the Sinopean amphora stamps may be relevant to this discussion. In 1998, Conovici published nearly 550 such stamps from Histria. The highest number of attributes in this sample, about 28%, refers directly or indirectly to wine (grapes, silens and satyrs, symposium vessels). The next highest incidence, about 26%, depict various gods or their attributes (Nike, Hekate, Hermes, Artemis, Herakles, Helios). Animals (mainly birds and lions) make up about 13%, and attributes related to ships and seafaring about 10%. Unidentified persons and other motifs (trophies, horns etc.) constitute about 10% each. Finally, 2% depict the emblem of the coinage of Sinope: an eagle clutching a dolphin.[23] True, we cannot be sure that the symbols on the stamps have anything to do with the contents of the amphorae, and this is in any case only a rough count. Still, it tends to support the notion that Sinopean amphorae were primarily containers of wine.[24] None of the attributes seems to relate to olive oil or grain, and the one with the eagle and a dolphin is probably emblematic for Sinope in the same way as a rose or the head of Helios were for Rhodes on Rhodian amphora stamps.

Thus, according to current scholarship the amphorae produced in the Black Sea region in the Classical and Hellenistic periods were primarily intended as containers of wine and to a lesser degree of olive oil and grain. No one since Stoddart has claimed that any amphora type was primarily manufactured to carry fish products, but Garlan has stated that it is "tempting to think that salted fish products could have been the main contents of the about 181

Sinopean amphorae found sporadically in the Mediterranean".[25] He also mentions olive oil and wine in connection with Sinopean amphorae,[26] and a recent find has given his proposal a new actuality.

3. *The Varna shipwreck*

In January 2003, there were reports in the international press that a joint Bulgarian-American expedition directed by Robert Ballard had discovered a shipwreck off Varna at the eastern coast of Bulgaria.[27]

The wreck contained at least 20-30 amphorae, but only one of these was retrieved, which allegedly "looked like a type of amphora that would be manufactured at the site in Sinop, Turkey." It is said that

> recent analysis of sediment gathered from inside the amphora revealed that it contained bones of a large freshwater catfish species, several olive pits, and resin ... Cut marks visible on the fish bones, together with other physical clues and references from classical literature, lead researchers to believe the amphora carried fish steaks—catfish that was butchered into six- to eight-centimeter ... chunks and perhaps salted and dried for preservation during shipping ... Radiocarbon analysis of fish bone samples taken from the amphora ... indicated that the bones were between 2,490 and 2,280 years old [i.e. between ca. 487 and 277 BC].

A more intensive investigation of the wreck will, hopefully, clarify whether we are dealing with "a big supply boat full of butchered fish" as Ballard tends to think, or if – as noted by maritime archaeologist Cheryl Ward: "the other amphorae weren't carrying fish, these fish may just have been somebody's lunch".[28] Perhaps new light will also be shed on the curious fact that olive pits and resin were found with the fish bones inside the amphora, which could suggest that the amphora had been re-used.

4. *The question of secondary use*

This leads to the question if there is evidence for a trade-related re-cycling of transport amphorae in the Classical and Hellenistic period.

This is a problematic topic, not least because of the difficulties involved in distinguishing between primary and secondary use, for instance in the case of charred remains of "bones and fish scales" found together with grains of wheat, barley, millet and lentil seeds in an amphora in a cellar at Olbia datable to the third to second century BC.[29] There is ample evidence of re-use of transport amphorae at a local level,[30] for instance as storage vessels. The amphorae found in Room 13 of the Monumental Building U6 at "Panskoye I" were thus re-used as containers of oil and/or grain.[31] *Dipinti* and *graffiti* are suggestive of such a re-use.[32] Re-cycled amphorae were used as building material, burial containers for infants, and in numerous other ways. However, there seems to be no evidence for a large scale systematic re-use of transport amphorae in inter-regional trade in the periods dealt with here.[33]

In order to credit that re-cycled amphorae played an important role in such an endeavour, one would have to presuppose the existence of a system for gathering containers after their use and transporting them to wine and olive presses or other production facilities. But there appears to be no evidence at all for this in the periods in question.[34]

5. *Literary evidence for trade in fish products from the Black Sea*

Still, it is pertinent to be reminded that "absence of evidence is no evidence of absence", because the ancient written sources document the importance of fishing in the Sea of Marmara and in the Black Sea.[35] Moreover, a number of texts mention a trade in fish and fish products in the Classical and Hellenistic Greek world.[36]

A reasonably well-preserved papyrus from the Zenon archive, for instance, records the valuation (for tax purposes) of goods imported to Egypt on two ships on behalf of Apollonius and others. This document, which dates from May-June 259 BC, lists among other goods "[- -] dried fish", "[- -] fish pickled in the season", "5 jars of [belly of tunny fish/ at 20 dr., [100 dr.]", "[- -] of salted fish at 16 dr.", "[- -] of mullet at dr. [- -]", immediately after which follows the entry "2 earthenware jars of wild boar-meat at 2 dr. [4 dr.]". True, the origin of these goods is not mentioned, and there is little reason to regard them as originating in the Black Sea region, even if 10 *choinikes* of Pontic nuts are mentioned further down the list.[37]

Another snippet of information comes from the fourth-century Demosthenic speech 35 *Against Lacritus*. The trial in which this speech was held concerned a maritime loan of 3,000 drachmas given to two merchants operating from Athens for a return voyage to the Pontus. The merchants had been contractually obliged to buy 3,000 amphorae of wine from Mende or Scione, which they were then to sell or exchange with goods from the Black Sea (Dem. 35.10-13, for the contract). However, on their return to Athens, the moneylender accused the merchants of having violated almost every single clause of the agreement. In particular, they had allegedly taken on board only 450 jars of wine (35.19) and had fabricated the loss of the return cargo in shipwreck (suffered en route from Pantikapaion to Theodosia) in order to explain the fact that they had come back to Athens empty-handed (35.31). It appears that the defendants had claimed that they were actually returning with a cargo to Athens, including salted fish (35.31). But this is met with the counter-claim that the consignment of fish was in reality transported from Pantikapaion to Theodosia on behalf of a certain farmer to be used by the workforce in his farm, and that, at any rate, it only amounted to a mere eleven or twelve jars, taken on board along with two hampers of wool and two or three bundles of goatskins (35.32, 34). Whatever the truth may have been about this matter, the merchants at least thought their claim that the Black Sea fish was destined for Athens would sound credible. The fact that a fragment from the work of a comic poet, preserved by Athenaios, informs us that the Athenians credited Pontos with producing the best salt-fish, suggests an import of this commodity on a certain scale.[38]

It is beyond the scope of this paper to present a systematic analysis of all the written sources. Such an endeavor might provide a clear answer to the question about the scale of the trade in fish and fish products, and perhaps also indicate if such an exchange was mainly conducted within the region itself or was part of a larger interregional trade. However, the instance cited from the Hellenistic period shows that ceramic containers – but not necessarily amphorae – could indeed be used for transporting fish, and there is nothing in either text to indicate that this was in any way unusual; the Zenon papyrus, in addition, documents that such vessels were used for meat too. At the same time we cannot exclude the simultaneous use of other, non-ceramic forms of transportation, for instance baskets.

6. *Conclusion*

The fragmentary evidence available suggests that the amphorae produced in the Black Sea region in the Classical and Hellenistic periods were not primarily intended as containers of fish or fish products. Still, the Varna wreck suggests that they could – at least occasionally – be used for this purpose. It is a moot question if the amphora in question carried supplies for the ship's crew or whether it was part of a larger consignment. But even if the latter turns out to have been the case, it is doubtful if a large-scale and regular trade in fish products could have been based on re-cycled amphorae. In that case, one would expect to find evidence of a systematic collecting of re-usable amphorae. Also, one would expect amphorae found in wrecks to present a less homogeneous picture than what is actually the case. It may therefore be tentatively concluded that such a trade was either at a small scale or irregular.[39] This accords well with T.W. Gallant's conclusion that fish from the Black Sea, which reached the Mediterranean "was a luxury item, aimed at a very restricted market. It cannot be taken as symptomatic of society as a whole (Polybius, 4.38.3-7)".[40]

There is precious little archaeological evidence to support the notion that the Black Sea region was the focus of a large-scale and systematic amphora-based trade in fish and/or fish products in the Archaic, Classical and Hellenistic periods.[41] The fact that stamped amphorae produced in the Black Sea region only occur sporadically south of the Bosporus certainly suggests that the scale of any such trade must have been restricted. No one has yet mapped the distribution of Black Sea amphorae in the Mediterranean, but among 1001 amphora stamps from Athens recently published by Gerhard Jöhrens, only six came from Sinope, and one from Chersonesos.[42] This situation seems typical, and only 181 Sinopean stamped amphorae are known from the entire Mediterranean.[43]

From an archaeological point of view, then, there is much to be said for Gallant's view that fish from the Black Sea region were luxury items, which "titillated the palates of discerning ancient gourmets".[44] However, it does not follow that the trade was of negligible economic import. The few written sources can hardly be construed as evidence of a large scale and systematic exportation of fish or fish products from the Black Sea to the Mediterranean in the period under review. Still, they tell us that Pontic salt fish were appreciated in Athens. Hence, it would be imprudent to make too much of the absence so far of any tangible evidence from the existing amphora record: other types of vessels than amphorae (possibly unrecognised by archaeologists) – not to speak of wooden crates, baskets or other containers of perishable materials – might have been involved in such a trade.

Notes

1 Rostovtzeff 1941, 1254.
2 Gallant 1985, 35.
3 Stoddart 1850, 98-99.
4 Stoddart 1853, 56-57.
5 For the Roman period, cf. Curtis 1991, 118-129.
6 Cf. Eiring and Lund (eds.) forthcoming.
7 Peacock and Williams 1985, 2.
8 Cf. Lund 2002 and forthcoming.
9 Will 2001, 263.
10 Koehler 1996, 326.
11 Garlan 2000, 90-91.
12 Lawall 1998; cf. also idem 1999.
13 Lawall 2000.
14 For an overview, cf. Garlan 2000, 195-96 and the bibliographical surveys Empereur and Garlan 1987, 1992, 1997 and Garlan 2002 with references to Russian scholarship.
15 Cf. Garlan (ed.) 1999.
16 Kac et al. 2002, 108.
17 Achmerov 1947, 175; Stolba 2002, 235.
18 Kac et al. 2002, 105.
19 Bittner 1998, 119-120 concludes that "bleibt der Weintransport in den Amphoren eine glaubwürdige, jedoch noch endgültig zu beweisende Arbeitshypothese"; cf. also Jefremow 2003, which deals with "Der Wein vom taurischen Chersonesos in Thrakien"; see further Lomtadze and Zhuravlev, forthcoming.
20 Conovici 1998, 169: "du vin surtout".
21 Kac et al. 2002, 108.
22 In the *Anabasis* (6.1.15), Xenophon mentions that the Sinopeans "sent to the Greeks, as gifts of hospitality, three thousand medimni of barley meal and fifteen

hundred jars of wine", whereas Strabon (13.3.12-13) states that the land around the mouth of the Halys was fertile, "productive of everything ... and planted with olive trees".

23 Conovici 1998, 193-194. Apparently no attributes refer to olives or olive oil.

24 Cf. Garlan 1990, 499-503 figs. 8-9 for a stamp showing the treading of grapes.

25 Garlan 2000, 89.

26 Garlan 2000, 88-89.

27 http://news.nationalgeographic.com/news/2003/01/0110_030113_blacksea.html.

28 http://www.nationalgeographic.com/events/releases/pr011603.html.

29 Pashkevich 2001, 515-16.

30 Cf. Kent 1953, 128; Grace 1962, 108-9; Lawall 1995, 19-20; Garlan 2000, 180, note 28; Jefremow 2002, 38, note 45.

31 Cf. Ščeglov 2002, 53-54; Stolba 2002, 232 ad H 13; 235 ad H 33.

32 Cf. Will 2001; Stolba 2002, 235; 237.

33 Cf. Lawall 1995, 19 note 14: "In the course of researching and writing this dissertation many people have suggested to me that amphoras may have been re-used in successive cargoes following their initial exportation from point A and importation at point B. The only evidence I can think of for such an event would be if one were to find amphoras of demonstrably different date on the same shipwreck. I know of no such occurrence, nor can I find any other evidence for such a possibility of reuse'.

34 Garlan 2000, 179: "D'autant qu'on peut aussi se demander si les amphores grecques n'étaient jamais commericalisées vides, pour elles-mêmes, comme simple contenants: on l'a parfois affirmé, à tort me semble-t-il". Cf. Dupont 2001, 454: "des livraisons de cargaisons entières d'amphores vides sont également envisageables à plus grande distance: Je n'en connais pas vraiment pour le monde grec". Jefremow 2003, 38 states that "die Zeugnisse aus Ägypten, Zeleia, Histria, Olbia und Delos sprechen einheitliche Sprache, dass die leeren Amphoren gesammelt und deponiert werden". However, the instanced quoted seem all to relate to reuse at a local level, and the author considers transportation of empty amphorae from Herakleia Pontike "weniger wahrscheinlich".

35 Cf. for instance Dumont 1976-1977 and Mehl 1987, 115-117 and passim.

36 Bekker-Nielsen 2002a and 2002b mainly discuss evidence from the Roman period but refer to the seminal publications dealing with the previous periods.

37 *P. Cairo Zen.* 59.012, col. II, lines 38-48 trans. Austin 1981: No. 237; the Greek word used is *keramion*.

38 Hermippos fr. 63 (*Poetae Comici Graeci*), cf. Garnsey 1999, 117.

39 It is hardly warranted to project conditions in the Roman Empire backwards in time, when several types of Spanish amphorae were specifically designed to carry garum and other fish products. Peacock and Williams 1985, 113-114 Class 14: "Tituli picti suggest fish-based products" (Dressel 12; Beltran III; Ostia LII), 122-123 Class 18: "Fish-based products" (Dressel 38; Beltran IIA; Ostia LXIII; Camulodunum 186C; Pélichet 46; Callender 6); 124-125 Class 19: "Fish-based products" (Beltrán IIB; Ostia LVII); 126-127 (Dressel 14; Beltrán IVA (Parker 1977); Ostia LXII); 128-129 Class 21: "Fish-based products are suggested" (Beltrán IVB (Parker 1977); Ostia LXI); 130-131 Class 22: "Fish-products have been suggested" (Almagro 50; Ostia VII; Keay 1984 XXII); 149-50 Class 31: "perhaps wine for the French vessels and wine or fish products for the Spanish" (Dressel 28); 153-54 Class 33: "generally thought to be fish products" (Africana I "Piccolo"; Beltrán

57; Ostia IV; Keay 1984 III); 155-57 Class 34: "fish products may also have been carried" (Africana II "Grande"; Beltrán 56; Ostia III; Keay IV-VII). Vladimir Stolba has kindly informed me that this is also supposed for one large-sized variety of the Bosporan jars found in Tanais.

40 Gallant 1985, 35 and *passim*. Cf. Bekker-Nielsen 2002b for a critical assessment of Gallant's views about net-fishing in the Roman world.

41 Tsetskhladze 1998, for instance, only discusses the evidence for trade in grain, metals and slaves.

42 Jöhrens 1999.

43 Garlan 2000, 89.

44 Gallant 1985, 35. Cf. further Curtis 1991, 126: "Italy was importing it [i.e. processed fish from the Black Sea] as early as the second century B.C. ... Its expense may indicate that commerce in the Western Mediterranean in salted fish from the Black Sea was not fully developed or that only the most expensive kinds came from this region". According to Waelkens et al. 2003, 60, "it is commonly assumed that in Greece it [i.e. fish] became a staple food, certainly among common people ... Only the Imperial period saw the emergence of fermented and salted fish products prepared in quantity and traded, especially among urban consumers."

Size Matters: Estimating Trade of Wine, Oil and Fish-Sauce From Amphorae in the First Century AD[1]

Bo Ejstrud

Garum – and the other types of fermented fish-sauce – was a popular part of the Roman cuisine. So important in fact that Tønnes Bekker-Nielsen argues, that although garum was not among the *big three* food items of the Roman world – wine, oil and grain – it was important enough to be a strong candidate for fourth place.[2] The arguments are based on the abundant fragments of amphorae used – mainly – to transport liquids, and found in settlements throughout the Roman world. Amphorae for garum make up a sizable part of the fragments, typically 10-20%, and in some cases even more.[3]

But while we may infer from the proportion of fragments to the proportion of *amphorae*, which suggests that somewhere between every 5 and 10 amphora on a given Roman site contained garum, there is an obvious problem if we use these numbers to describe the proportion of *volumes*. Even disregarding the complex problems of taphonomy, there is the simpler problem of size: Every time we find a typical Dressel 20 amphora, or fragments of it, we have found 60-70 litres of olive-oil, while the average Dressel 7 amphora contained no more than 14-18 litres of garum. A simple count of sherds will overestimate the volume of garum to oil, in this example by a nominal factor of four (*cf*. Fig. 1).

Size *does* matter in a realistic assessment of trade. An estimation of the volumes cannot rely on the number of sherds, but rather, the numbers must be weighted against the volume of the containers.

This is not a simple matter, given the vast typological variation of amphorae. An estimation of volumes requires knowledge not only of the mean volume of each type, but also of the specific composition of types on individual sites. This again requires extensive excavations, and also detailed publications of the pottery from the sites excavated.

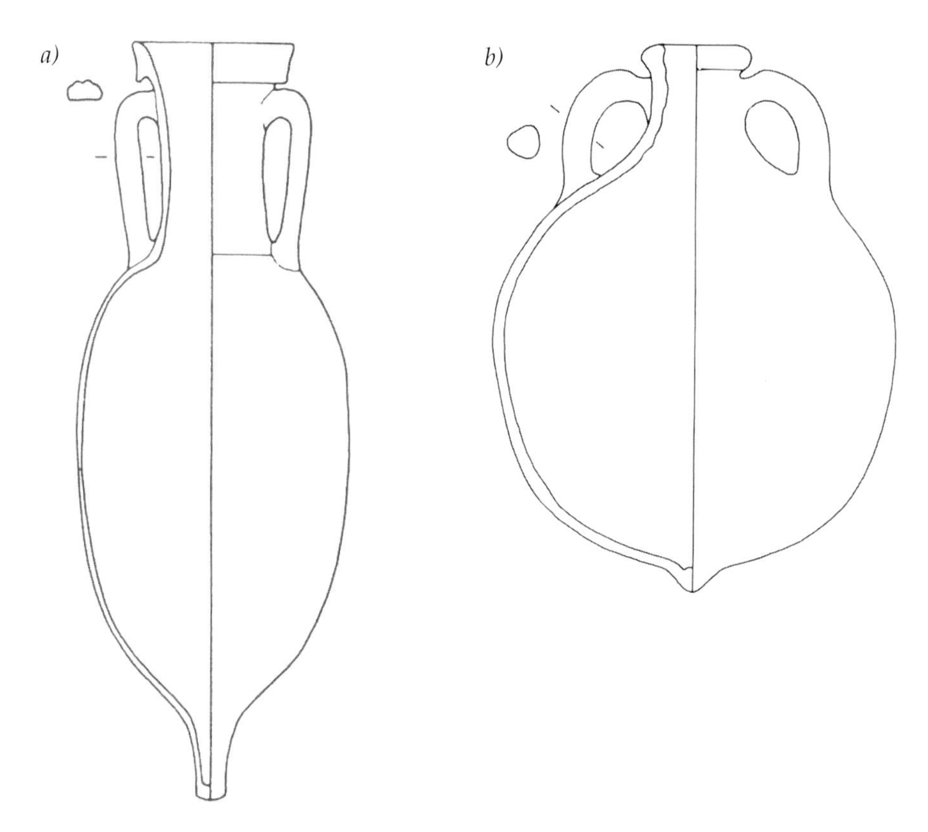

Fig. 1a-b. a) A Dressel 7 containing an average of 16 litres of garum (after Martin-Kilcher 1994a, 394). b) A Dressel 20 containing an average of 66 litres of oil (after Martin-Kilcher 1987, 55).

1. *The amphorae from Augst*

Well-published sites are in short supply in archaeology. But one such is the Roman colony of Augusta Raurica, today's Augst, lying at the Rhine some 10 km west of Basel in Switzerland (Fig. 2). The colony was founded in the mid-first century BC, and developed into a town of some 20,000 inhabitants during the second and early third century AD.

The extensive excavations are published in a long series of monographs – the series *Forschungen in Augst* having reached its thirty-second volume (Berger 2002) – along with numerous scientific and popular articles. The latest available bibliography (Anonymous 2001) lists more than 800 titles with Augst as the main theme.

Stephanie Martin-Kilcher published the amphorae from Augst in three extensive volumes (1987, 1994a, 1994b). The catalogue (Martin-Kilcher 1987, 1994b) lists almost 6000 numbers of typologically identifiable amphora-sherds, and apart from the typological analysis and presentation of the find, she also

Fig. 2. The colony of Augusta Raurica. General plan with the Rhine, important buildings and roads (redrawn from Martin-Kilcher 1987, 17).

discusses central aspects of trade in Augst, which all had to rely on imports, as neither oil, wine nor garum were produced locally.

1.1 The database

For the purpose of publication a database was originally developed to register the amphorae. Unfortunately this database is not published in electronic format, the only accessible traces of it being the lists forming the printed catalogue. Had the database been available however, for instance on a disk inlaid in one of the volumes, it is very doubtful whether it would be readable by any standard programme today. But the advantages of using electronic registration are so obvious, that the database was reconstructed for this purpose, although in a spreadsheet format, to facilitate calculations.

Over time there are considerable fluctuations in the relative number of types present at Augst.[4] To get a more stable picture the following will focus on finds from AD 30-70, which is the best represented. Not all finds can be dated precisely within this time frame, but following the method of Martin-Kilcher,[5] sherds are counted with a proportion equivalent to the number of years their context falls within the timeframe. A sherd found in layers that is dated to AD 10-50 counts with 0.5, as half the years are within the period. Assuming uniform distribution, this method should give a reasonable estimate, although the resulting sums are not integers. Sherds that cannot be dated within 100 years are disregarded.

The sherds are given both in total and minimum numbers, the latter being an empirical estimate reflecting the fact that any one pot can break into many pieces, and mainly counting the number of rims. As also shown by Martin-Kilcher,[6] there are no significant differences in the relative proportion between minimum and total number of sherds, so counting either way has no effect on the results in this case: Roughly 50% of the total sherds is part of the minimum number.

The stable relation between total and minimum numbers is reassuring in terms of taphonomy. With the total numbers we get the fragments that can be identified as belonging to any one type. With the minimum numbers, we get the number of identifiable fragments that can be separated from each other. Since the relation between what can be identified and what can be separated is much the same for all types, we get a good indication of the representtativity of the material: The larger amphorae do not seem to break into more – identifiable – pieces than the smaller ones, nor are they easier to identify. This is largely a product of the fact that the vessels are mainly identified by elements such as rims and handles, which do not vary in numbers with the size of the amphora.[7]

Establishing the new database was a simple matter of setting up various functions in the spreadsheet, and reading through the 6,000 numbers in the catalogue. More difficult was to determine the mean volume of the individual types. For the more abundant types such numbers can be found in the literature; Martin-Kilcher provides some, while another important source has been Paul Thyers very impressive web-based *Atlas of Roman Pottery*, concentrating on British sites. Most types are found in this way. For the more rare types, or those so far only known in fragments, there are no published estimates of their volumes. But for a majority of these problematic types, the Martin-Kilcher publication provides reconstructed drawings in a scale of 1:10. Using a GIS,

			All		30-70	
Type	**Content**	**µ(litre)**	**MIN**	**Total**	**MIN**	**Total**
Augst 17	Garum	20	65	98	9.4	17.8
Dressel 12	Garum	20	1	1	0.0	0.0
Dressel 7, 10, 11	Garum	16	48	55	20.2	24.9
Dressel 8	Garum	21	26	47	10.4	10.9
Dressel 9	Garum	42	27	38	7.8	9.8
Vindonissa 586	Garum	19	36	41	20.3	21.1
Pélichet 46	Garum	32	178	213	42.7	49.8
Augst 28	Garum	26	1	1	0.0	0.0
Augst 29	Garum	17.5	1	1	0.0	0.0
Augst 30	Garum	26	53	116	0.0	0.0
Group 9 ??	Garum	28.1	10	265	0.0	94.0
Dressel 10 sim.	Garum	19	49	70	20.9	27.0
Dressel 9 sim.	Garum	30	297	486	91.3	148.5
Group 10 ??	Garum	28.6	0	115	0.0	28.6
Augst 33	Garum	42	33	48	7.2	9.2
Augst 34	Garum	20	6	8	0.5	0.5
Augst 35	Garum	15	2	4	1.0	2.3
Dressel 16	Garum	10	4	6	0.4	1.2
Dressel 6A	Garum	26	1	1	0.3	0.3
Dressel 14	Garum	26	1	2	0.0	0.0
Dressel 20	Oil	66	974	2,009	265.5	474.6
Dressel 6B	Oil	20	2	3	1.3	1.7
Vindonissa 592	Oil	20	10	18	5.3	9.0
Augst 63	Oil	60	8	9	1.0	2.0
Tripolitana 1	Oil	55	2	4	2.0	3.5
Dressel 1	Wine	26	1	5	1.0	1.7
Pascual 1	Wine	25	12	21	4.1	5.4
Dressel 2-5	Wine	28	118	316	38.8	104.3
Camulodunum 184	Wine	17.5	29	121	10.2	35.6
Dressel 43	Wine	17.5	2	4	1.5	2.1
Oberaden 74	Wine	30	2	2	1.0	1.0
Gauloise 1-5	Wine	31	617	1172	88.8	169.7
Camulodunum 139	Wine	23	32	63	3.2	6.1
Augst 55	Wine	6.5	1	5	0.0	0.0
		Sum	*2,619*	*5,368*	*656.1*	*1,262.3*

Table 1. Amphorae from Augst. Types not known in first-century context are not shown.

it is possible to digitalize these drawings in true scale, and then measure the area of the cross-section of each type. By comparing these areas to amphorae of known size, it was possible to get an estimate of the volumes. The character of guesstimate must still be stressed for some of the types, but especially those very rare, and consequently with little influence on the results. The volume of the undetermined sherds of Martin-Kilcher group 9 and 10 are set as weighted averages.

1.2 *Relative imports*

With these numbers established (Table 1), the amphorae from Augst provide a good example of the difference between counting sherds and measuring volumes. According to the simple count of the pottery, the imports of wine, oil and garum are roughly comparable in size, with 1/3 each (Fig. 3, left). But weighting the occurrence of individual types with their volume gives a very different picture (Fig. 3, right). Wine is reduced to a mere 15% of the total volume while oil is the dominant product, with almost 2/3 of the total. Figure 3 should prove the point of this paper's title.

1.3 *Absolute imports*

The relative abundance of sherds is relatively unproblematic to establish. There are some obvious problems of statistical and taphonomic character, and probably some less obvious ones too, but with excavations at this scale the result can be considered statistically relatively stable.

Estimating the absolute volumes is much more problematic. Acknowledging this, but also pointing to the fact that the number of fragments actually found equivalate an annual import of only 2½ amphorae of oil, or 165 litres, Martin-Kilcher gives an estimate of the imports of oil.[8] Several historical and archaeological sources are used to estimate an annual consumption of nine Roman pounds of oil per capita. Based on an average population of 12,000 people, and assuming that only half the population were actual consumers, the result is an import of 270 amphorae per year, or 17,820 litres. The estimate seems to be on the conservative side, but gives us an idea of the actual volumes

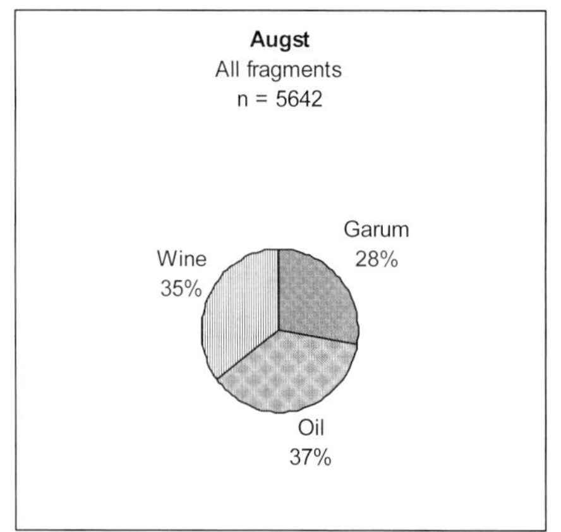

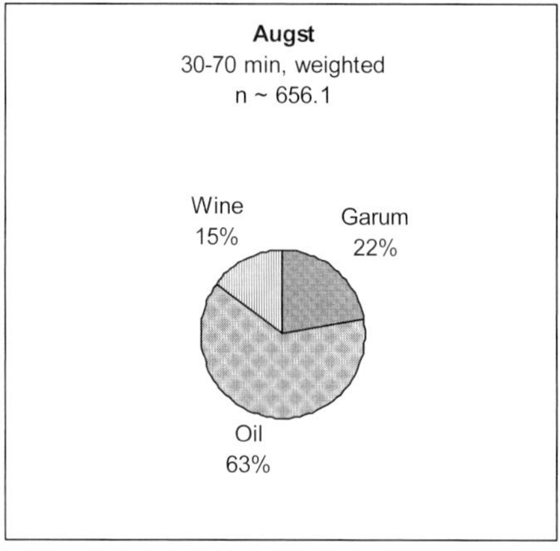

Fig. 3. Wine, oil and garum at Augst. Left: The proportion between all fragments (Martin-Kilcher, Abb. 204). Right: The proportion between volumes at AD 30-70.

consumed. Using this estimate as a basis, the annual import can be calculated as roughly 4,000 litres of wine and 6,000 litres of garum (Table 2).

It is difficult to relate such numbers to anything real. But using the same basis as Martin-Kilcher; a population of 12,000 of which only half are actual consumers, these numbers would be equivalent to a consumption of 0.7 litres – or one bottle – of wine per person per year. Where in relative numbers the low representation of wine in Augst was remarkable, these numbers tell us that there must be something wrong, and that even the meticulous counting and weighting of each individual container gives us a false picture of the actual imports. Size is not enough to get a reasonable estimate.

2. *Trade in oil, wine and garum*

One way to explain this problem is to look at the distribution at other contemporary Roman sites. In fact a detailed investigation would require all such sites to be as well published as Augst, and the establishment of new detailed databases of every fragment of pottery. But if we keep the investigation within Western Europe, and within the first century AD, it should be possible to apply the Law of Averages, using the average volume of amphorae from Augst to estimate the volumes at other sites. These average volumes are given in table 3.

	Volume (litres)
Wine	4,285
Oil	17,820
Garum	6,271

Table 2. Estimated annual imports of wine, oil and garum in Augst.

	Volume (litres)
Wine	29.0
Oil	65.5
Garum	27.9

Table 3. Mean volumes of amphorae at Augst.

Martin-Kilcher (1994a) provides numbers for the relative occurrence of sherds at other sites as well. Limiting the investigation to the first century AD and omitting the villas, which are not comparable to the towns, we have the relative composition of sherds from wine, oil and garum from another four sites: Nijmegen, Avenches, Saint-Romain-en-Gal and Rome. These are not many to compare with, but they provide a cross-section of Europe from Nijmegen in the North to Rome in the South. Weighting the relative numbers of fragments from each site by the average volumes from Augst, these few sites actually outline a distinct geographical pattern (Fig. 4).

The strong dominance of oil is not particular to Augst but can also be seen at Nijmegen and Avenches, all three situated in the northern part of the Empire, while the two southern, but otherwise very different, sites of Saint-

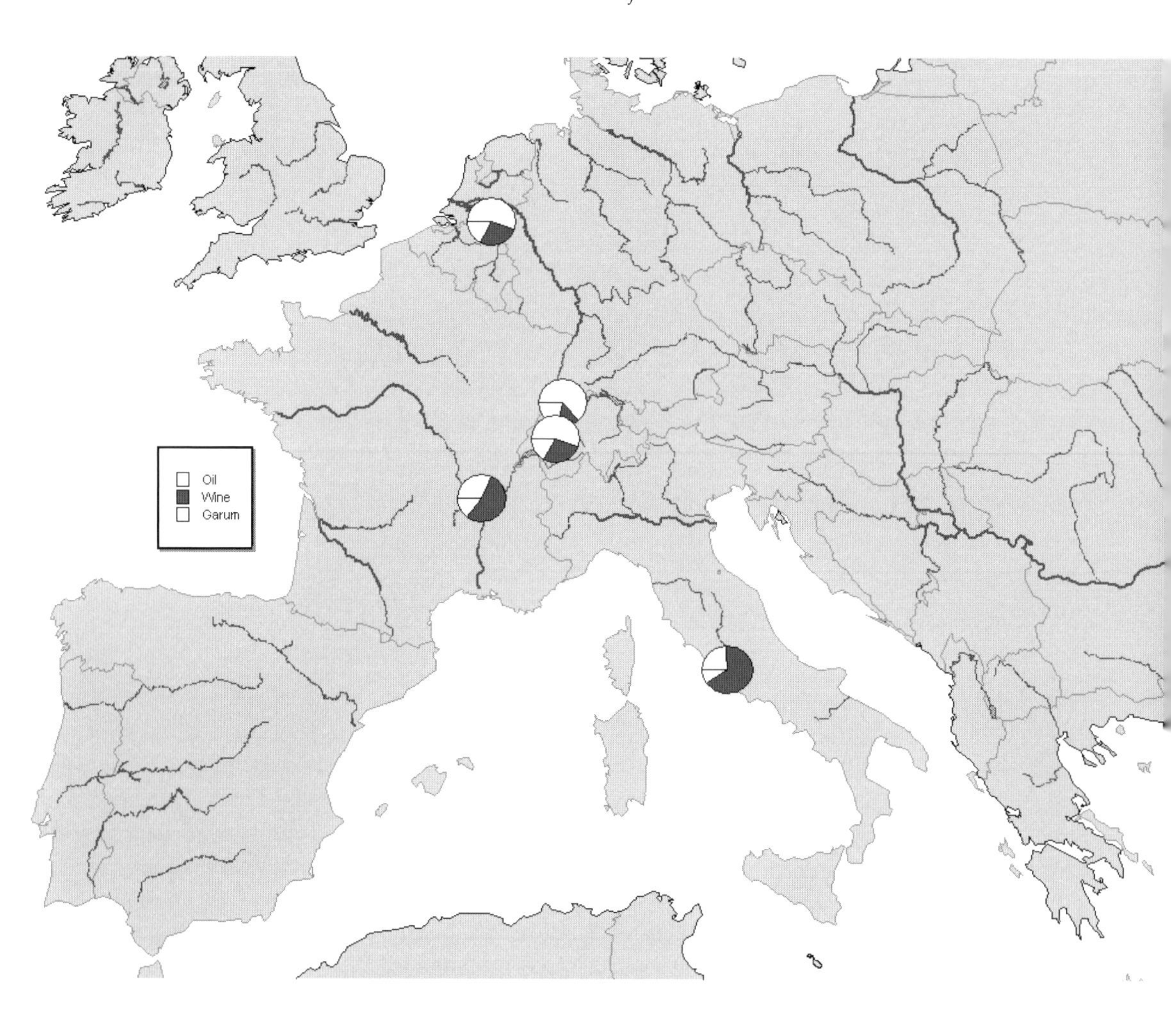

Fig. 4. Relative volumes of oil, wine and garum in Western Europe. Modern borders from World Data bank 2 (CIA 1972).

Romain-en-Gal and Rome are both dominated by wine. Apparently there are two separate groups in this material, with oil-dominance in the north, and wine-dominance in the south.

One explanation could be that wine in the northern part of the Empire was primarily transported in barrels, and is not represented with the amphorae. This explains the dominance of oil in the North. Since barrels are difficult to document archaeologically, wine becomes invisible to us. This explanation can be corroborated by the fact that the relation between oil and garum is very stable between all of the sites, with 25-30% of the volumes being garum (Fig. 5). Although figure 4 seems to indicate two very different patterns of consumption, figure 5 points to the fact that wine is the variable factor.

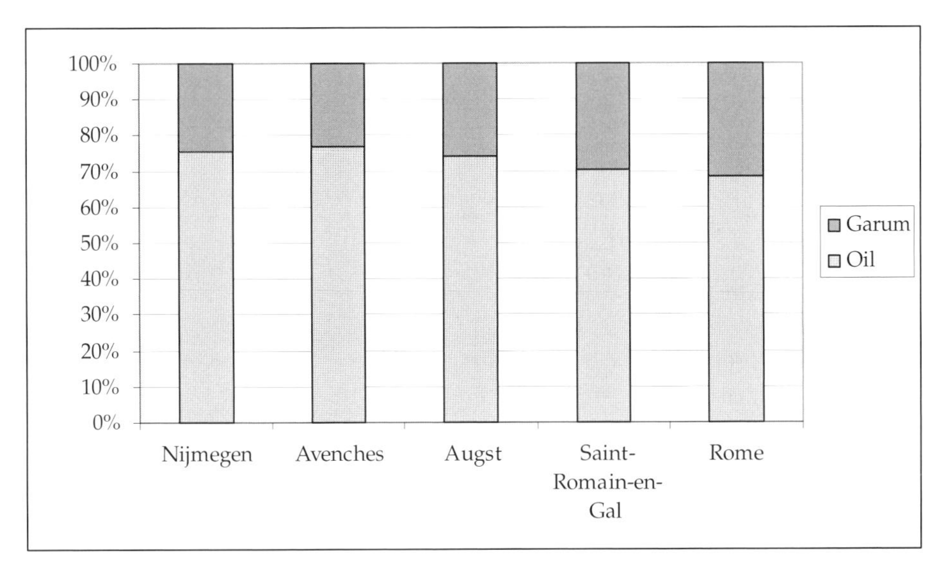

Fig. 5. The relation between oil and garum on the five sites.

During the first century AD both oil and garum were produced around the Mediterranean, strongly dominated by Andalusia in Southern Spain. These products came in amphorae, and they are therefore relatively easy to find, which explains the stable relation found on Fig. 5. Opposite this, there was a local production of wine in Gaul, which was mainly distributed in barrels, making wine more difficult to find.

Assuming that the patterns of consumption were much the same across Europe, one can speculate as to how much wine is missing in Augst. The average volumes on the two Mediterranean sites in this investigation are 62% wine, 28% oil and 10% garum. These averages can be replicated in Augst, if we assume that *c.* 90% of all wine came to this site in barrels, thus being invisible in the archaeological record. Using the absolute numbers calculated by Martin-Kilcher, where 17,820 litres of oil were consumed every year, this

	Mediterranean	**Augst**	
	Relative volumes, %	Absolute, litres	Relative volumes %
Wine	62.25	39,725	62.25
Oil	27.92	17,820	27.92
Garum	9.83	6,271	9.83

Table 4. The average imports to Saint-Romain-en-Gal and Rome can be replicated exactly in Augst by simply changing the volumes of wine. Cf. table 2.

would equal an "invisible" annual import of *c.* 35,500 litres of wine in barrels (Table 4).

The absolute volumes only serve as illustration. What is important is to point to the high degree of stability in the relation between oil and garum. They are specialized products, produced in distinct regions and distributed in containers, which make them easy to trace. Apparently they were also imported in equal proportions across the Empire – or at least the western part of it, which is examined here.

3. *Domestic imperialism – towards a conclusion*

The Roman way of life left us with some very spectacular features. The villas, aqueducts, baths and arenas are well known elements across the Empire, reflecting that Roman soldiers and administrators brought their culture with them, but also that the local populations were "Romanised", sometimes remarkably fast. These spectacular buildings are an important and highly treasured part of the local cultural heritage wherever the Romans went.

What this study seems to suggest is that "Romanisation" is not just about the spectacular. It is also reflected in the very near and personal question of what to eat. This is not a matter of whether the Roman and Romanised people utilised local food resources. They did, but what is suggested by figure 5 is that there was very little regional variation in how much oil you would cook your food in, or how much garum you would pour over it to get that genuine Roman flavour. Romanisation happened not only in the public sphere of gladiatorial games and water supply. It was also a domestic matter with a tangible impact on the dinner table. Speaking of a "domestic imperialism" is not misleading altogether.

If this interpretation is true, it also has implications on the discussions of substitution, i.e. whether oil can be replaced by animal fat, wine by beer – or garum perhaps with salt.[9] Before going into this discussion it is important to recognize that in working with imported goods such as wine, oil and garum, our data mainly reflect the situation of the upper social strata. Martin-Kilcher assumes that only half the population in Augst had access to the imported olive-oil.[10] The remainder of the population, whatever fraction it may have been, must have used another type of fat in their food, and had no choice but to find a substitute for oil.

But for those who had the means there are no real suggestions of such a substitution. The relative consumption of both oil and garum is the same from the Netherlands to Italy, demonstrating that those who could buy these foods also did. Wine and its possible substitution with beer are more difficult to assess as neither can be documented archaeologically. But following the line of reasoning above, that Spanish oil and garum was brought all across Europe because it was a proper part of Roman lifestyle, one could speculate that beer – with which Tacitus, at least, did not feel at ease (*Germania* 23.1)

– is an unlikely candidate to fill the barrels that are so obviously missing in both Augst, Avenches and Nijmegen.

Finally it is also important to remember that what is discussed here is based solely on sources from Western Europe. With the Black Sea Centre we have a unique opportunity for comparison with conditions at the other end of the Roman Empire. Just like those in Western Europe, the Romanised people at the Black Sea had villas, arenas and aqueducts. These features are known, and have been so for a long time. Maybe an interesting next step in the investigation of Roman impact and integration of the Black Sea would be to ask just how much garum they consumed.

Notes

1 I would like to thank the organizers for the invitation to speak at the workshop in Esbjerg. Dr. Tønnes Bekker-Nielsen is thanked both for the encouragement to make this study on the other side of the borderline between prehistoric and classical archaeology and for bibliographical help.
2 Bekker-Nielsen 2002a, 35.
3 Bekker-Nielsen 2002a, 34f.
4 Martin-Kilcher 1994a, 466ff.
5 Martin-Kilcher 1994a, 466.
6 Martin-Kilcher 1994a, 555.
7 Bekker-Nielsen 2002a.
8 Martin-Kilcher 1987, 193ff.
9 Cf. Bekker-Nielsen 2002a.
10 Martin-Kilcher 1987, 193ff.

Abbreviations

ACFM	Advisory Committee on Fishery Management
AE	*L'Année Épigraphique*
AJA	*American Journal of Archaeology*
ANRW	Temporini, H. & W. Haase (eds.), *Aufstieg und Niedergang der römischen Welt*
Asbor	*Archeologičeskij Sbornik Gosudarstvennogo Ermitaža*
BAR	British Archaeological Reports
BCH	*Bulletin de Correspondence Hellénique*
C&M	*Classica et Mediaevalia. Revue danoise de philologie et d'histoire*
CAF	Kock, T. (ed.), *Comicorum Atticorum Fragmenta*
CIL	*Corpus Inscriptionum Latinarum*
FAO	United Nations Food and Agricultural Organization
IAK	*Izvestija imperatorskoj Archeologičeskoj Komissii*
ICES	International Council for the Exploration of the Seas
IGBul	*Inscriptiones Graecae in Bulgaria Repertae*
IGRR	Cagnat, R. (ed.), *Inscriptiones Graecae ad Res Romanas Pertinentes*
IK	*Inschriften griechischer Städte aus Kleinasien*
JPrehistRel	*Journal of Prehistoric Religion*
JRA	*Journal of Roman Archaeology*
KSIA	*Kratkie Soobščenija Instituta Archaeologii AN SSSR*
MEFRA	*Mélanges de l'École Française de Rome. Antiquité.* Paris
MIA	*Materialy i issledovanija po archaeologii SSSR*
MünstBeitr	*Münstersche Beiträge zur antiken Handelsgeschichte*
NA IA NANU	*Naučnyj archiv Instituta archaeologii Nacional'noj Akademii nauk Ukrainy*
P.Wisc.	Sijpesteijn, P.E. (ed.), *The Wisconsin Papyri.* Leiden
RE	Pauly, A., G. Wissowa & W. Kroll (eds.), *Real-encyclopaedie der classischen Altertumswissenschaft*
REG	*Revue des Études Grecques*
RosA	*Rossijskaja Archaeologija*
SNG	*Sylloge Nummorum Graecorum*
SovA	*Sovetskaja Archaeologija*
VDI	*Vestnik drevnej istorij*
ZPE	*Zeitschrift für Papyrologie und Epigraphik*

Bibliography

Abaev, V.I. 1949. *Osetinskij jazyk i fol'klor*. I. *Skifskij jazyk*. Moscow.

Abaev, V.I. 1973. *Istoriko-etimologičeskij slovar' osetinskogo jazyka*. Leningrad.

Achmerov, P.B. 1947. Amfory drevnegrečeskogo Chersonesa, *VDI* 1947:1, 160-176.

Alekseev, A.Ju. 1995. Skifskoe pogrebenie V v. do n.e. v kurgane Malaja Cimbalka (raskopki I.E. Zabelina v 1868 g.), *ASbor* 32, 53-59.

Alexandrescu Vianu, M. 1997. Aphrodites orientales dans le bassin du Pont-Euxin, *BCH* 121, 15-32.

Andrusov, N.I. & S.A. Zernov 1914. Černoe more, in: *Krym. Putevoditel'*. Simferopol' (http://www.moscow-crimea.ru/atlas/more3.html).

Anochin, V.A. 1977. *Monetnoe delo Chersonesa IV v. do n.e. – XII v. n.e.* Kiev.

Anochin, V.A. 1986. *Monetnoe delo Bospora*. Kiev.

Anochin, V.A. 1988. Monetnoe delo i denežnoe obraščenie Kerkinitidy (po materialam raskopok 1980-1982 gg.), in: V.A. Anochin (ed.), *Antičnye drevnosti Severnogo Pričernomor'ja*. Kiev, 133-148.

Anochin, V.A. 1989. *Monety antičnych gorodov Severo-Zapadnogo Pričernomor'ja*. Kiev.

Anonymous 2001. Bibliographie / bibliography. Augst und Kaiseraugst http://www.baselland.ch/docs/kultur/augustaraurica/publ/bibliogr/1_bibliogr.htm. Updated September 2001. Cited February 2003.

Aquerreta, Y., I. Astiasarán & J. Bello. 2001. Use of Exogenous Enzymes to Elaborate the Roman Fish Sauce "Garum", *Journal of the Science of Food and Agriculture* 82, 107-112.

Aranegui Gascó, C. (ed.) 2001. *Lixus: colonia fenicia y ciudad púnico-mauritana, anotaciones sobre su occupación medieval*. Valencia.

Aruz, J., A. Farkas, A. Alekseev & E. Korolkova (eds.) 2000. *The golden deer of Eurasia. Scythian and Sarmatian treasures from the Russian steppes*. New Haven.

Aston, M. (ed.) 1988. *Medieval Fish, Fisheries and Fishponds in England*, I. Oxford.

Aubet, M.E. 1987. Notas sobre le economía de los asentamientos fenicios del sur de España, *Dialoghi di Archeologia* 5.2, 51-62.

Aubet, M.E. 1993. *The Phoenicians and the West*. Cambridge.

Aubet Semmler, M.E. 2002. The Tartessian Orientalizing Period, in: Bierling (ed.) 2002, 199-224.

Auriemma, R. 1997. Le anfore africane del relitto di Grado. Contributo allo studio delle prime produzioni tunisine e del commercio di salse ed conserve di pesce. *Archeologia subacquea.* (Studi, Richerche e Documenti, II). Rome, 129-155.

Auriemma, R. 2000. Le anfore del relitto di Grado e il loro contento. *MEFRA* 112, 27-51.

Austin, M.M. 1981. *The Hellenistic world from Alexander to the Roman conquest. A selection of ancient sources in translation*. Cambridge.

Badham, C.D. 1854. *Prose Halieutics or Ancient and Modern Fish Tattle*. London.

Barnes, T.D. 1998. *Ammianus Marcellinus and the Representation of Historical Reality*. Ithaca.

Basch, L. 1987. *Le musée imaginaire de la marine antique*. Athens.

Beddows, C.G. 1985. Fermented Fish and Fish Products. *Microbiology of Fermented Foods*. 2 Vols. London, Vol. 2, 1-39.

Bekker-Nielsen, T. 2002a. Fish in the Ancient Economy, in K. Ascani et al. (eds.), *Ancient History Matters. Studies presented to Jens Erik Skydsgaard on His Seventieth Birthday* (Analecta Romana Instituti Danici, Supplementum, 30). Rome, 29-37.

Bekker-Nielsen, T. 2002b. Nets, Boats and Fishing in the Roman World, *C&M* 53, 215-233.

Belon, P. 1555. *Les observations de plusiers singularitez de choses memorables, trouvées en Grece, Asie, Iudée, Egypte, Arabie, et autres pays estranges*. Antwerp.

Belov, G.D. & S.F. Strželeckij 1953. Otčety (Raskopki 1937-1948 gg.) Kvartaly XV i XVI, *MIA* 34, 32-236.

Belov, G.D. 1936. *Raskopki Chersonesa v 1934 g*. Simferopol'.

Belov, G.D. 1953. Severnyj pribrežnyj rajon Chersonesa (po novejšim raskopkam), *MIA* 34, 11-31.

Ben Lazreg, N., M. Bonifay, A. Drine & P. Trousset 1995. Production et commercialisation des *salsamenta* de l'Afrique ancienne, in: *Productions et exportations Africaines. Actualitiés archéologiques*. Paris, 103-142.

Berger, L. 2002. *Durchbrochene Messerfutterale (Thekenbeschläge) aus Augusta Raurica. Ein Beitrag zur provinzialrömischen Ornamentik* (Forschungen in Augst, 32). Augst.

Bernal Casasola, B. & J.M. Pérez Rivera 1999. *Un viaje discrónico por la historia de Ceuta; resultados de las Intervenciones arqueológicas en el Paseo de las Palmeras*. Madrid.

Bertier, J. 1972. *Mnésithée et Dieuchès*. Leiden.

Bierling, M.R. (ed.) 2002. *The Phoenicians in Spain: An Archaeological Review of the Eighth-Sixth Centuries B.C.E.* Winona Lake, Indiana.

Bilde, P. 1990. Atargatis/Dea Syria: Hellenization of her cult in the Hellenistic-Roman period? in: P. Bilde et al. (eds.), *Religion and religious practice in the Seleucid kingdom* (Studies in Hellenistic civilization, 1). Aarhus, 151-87.

Bittner, A. 1998. *Gesellschaft und Wirtschaft in Herakleia Pontike. Eine Polis zwischen Tyrannis und Selbstverwaltung* (Asia Minor Studien, 30). Bonn.

Blacker, L.V.S. 1922. *On Secret Patrol in High Asia*. London.

Blagg, T.F.C., R.F.J. Jones & S.J. Keay (eds.) 1984. *Papers in Iberian Archaeology*, II. Oxford.

Blanchard-Lemée, M. 1975. *Les mosaïques du quartier central de Djemila (Cuicul)*. Aix-en-Provence.

Blázquez, J.M., M. Montenegro, J.M. Roldán, J. Mangas, R. Teja, J.J. Sayas, L.G. Iglesias & J. Arce 1978. *Historia de España Antigua*, II: *Hispania Romana*. Madrid.

Blümner, H. 1869. *Die gewerbliche Thätigkeit der Völker des klassischen Alterthums*. Leipzig.

Blümner, H. 1911. *Die römischen Privataltertümer*. 3rd ed. (Handbuch der klassichen Altertumswissenschaft, 4,2,2). Munich.

Bode, Mathias 2002. Wale und Walfang in der Antike, *Laverna* 13, 1-23.

Bodjanskyj, O.V. 1951. Ščodennik archeologičnyh doslidiv na pravomu berezi Dnipra, *NA IA NANU*. 1951/12 (Manuscript).

Boessneck, J. (ed.) 1973. *Tierknochen von westphönizischen und phönizisch beeinflussten Ansiedlungen in Südspanischen Küstengebiet*. Munich.

Bonneville, J.-N., F. Didierjean, N. Dupre, P. Jacob, J. Lancha, M. Fincker, C. Ney & J.-L. Paillet 1984. La dix-huitiéme campagne de fouilles de la Casa de Velazquez à Belo en 1983 (Bolonia, province de Cadix), *Mélanges de la Casa de Velazquez* 20, 439-486.

Boplan, G.L. 1990. *Opys Ukrainy*. Kiev.

Borisov, A.A. 1956. O kolebanijach klimata Kryma za istoričeskoe vremja, *Izvestija Vsesojuznogo geografičeskogo obščestva* 88.6, 532-541.

Brabič, V.M. 1964. Ob izobraženii na monetach Pantikapeja l'vinoj golovy i osetra, *Soobščenija Gosudarstvennogo Ermitaža* 25, 50-52.

Brašinskij, I.B. 1968. Recherches soviétiques sur les monuments antiques des régions de la Mer Noire, *Eirene* 7, 81-118.

Braund, D. & J. Wilkins (eds.) 2000. *Athenaeus and his World*. Exeter.

Bravo Perez, J. 1980. Fábrica de salazones en la Ceuta romana, *CRIS Revista de la mar* April, 40.

Brujako, I.V. 1999. *Očerki ekonomičeskoj istorii naselenija Severo-Zapadnogo Pričernomor'ja v 7-3 vv. do R.Ch*. Volžsk.

Brun, A.H. 1930. *Blandt Krigsfanger i Turkestan*. Copenhagen.

Bruschi, T. & B. Wilkens. 1996. Conserves de poisson à partir de quatre amphores romaines, *Archaeofauna* 5, 165-169.

Bučinskij, I.E. 1953. Izmenilsja li klimat Ukrainy za istoričeskoe vremja, *Izvestija Vsesojuznogo geografičeskogo obščestva* 1, 21-30.

Buračkov, P.O. 1881. Opyt soglašenija otkrytoj v Chersonese nadpisi s prirodoju mestnosti i sochranivšimisja u drevnich pisatelej svedenijami, otnosjaščimisja ko vremeni vojn Diofanta, polkovodca Mithridata so skifami, *ZOOID* 12, 222-248.

Buračkov, P.O. 1884. *Obščij katalog monet, prinadležaščich ellinskim kolonijam, suščestvovavšim v drevnosti na severnom beregu Černogo morja, v predelach nynešnej Rossii*. I. Odessa.

Burdak, V.D. & A.N. Ščeglov 1966. O tempe rosta, vozrastnom sostave stad i migracijach nekotorych černomorskich ryb v antičnuju epochu, in: *Ekologo-morfologičeskie issledovanija nektonnych životnych*. Kiev, 117-120.

Burdak, V.D. 1966. Ob izmenenii tempa rosta černomorskich kefalej v istoričeskoe vremja, *Doklady Akademii nauk SSSR* 167.5, 1156-1158.

Cara Barrionuevo, L., J. Cara Rodríguez & J.M. Rodríguez López 1988. Las cuevas de la Reserva (Roquetas) y otras factorías pesqueras de época romana en la provincia de Almería, in: Ripoll Perelló (ed.) 1988, 919-934.

Carreras Monfort, C. 2000. *Economía de la Britannia Romana: la importacción de alimentos* (Colleccío Instrumenta, 8). Barcelona.

Casteel, R.W. 1976. *Fish Remains in Archaeology and Paleo-environmental Studies*. London.

Cepkin, E.A. 1970. Novye materialy k istorii rybnogo promysla v Tanaise, *KSIA* 124, 115-17.

Chibnall, M. 1975. Pliny's Natural History and the Middle Ages, in: T.A. Dorey (ed.), *Empire and Aftermath: Silver Latin II*. London, 57-78.

CIA 1972. *World Data bank 2*. Central Intelligence Agency, Washington, D.C. USA. GIS-layer downloaded from UNEP-GRID at http://www.grid.unep.ch/data/grid/gnv19.html.

Clément, V. 1999. Le territoire du Sud-Ouest de la péninsule Ibérique à l'époque romaine; du concept au modèle d'organisation de l'espace, in: Gorges & Rodríguez Martín (eds.) 1999, 109-120.

Cleto, J. 1995-96. A indústria de conserva de peixe no Portugal romano. O caso de Angeiras, *Matesinus* 1-2, 23-45.

Colls, D., R. Étienne, R. Lequément, B. Liou & F. Mayet 1977. *L'épave Port-Vendres II et le commerce de la Bétique à l'époque de Claude* (Archaeonautica, 1). Paris.

Conovici, N. 1998. *Histria VIII: Les timbres amphoriques, 2*. Sinope. Bucharest.

Corcoran, T.H., 1957. The Roman Fishing Industry of the Late Republic and Early Empire. Ph.D. dissertation, Northwestern University.

Cotton, H., O. Lerenau & Y. Goren. 1996. Fish Sauces from Herodian Masada. *JRA* 9, 223-238.

Coull, J.R. 1993. *World fisheries resources*. London.

Curtis, R.I. 1979. The Garum Shop of Pompeii, *Cronache Pompeiane* 5, 5-23.

Curtis, R.I. 1983. In Defense of Garum, *CJ* 78, 232-240.

Curtis, R.I. 1984a. *Negotiatores Allecarii* and the Herring, *Phoenix* 38, 147-158.

Curtis, R.I. 1984b. A Personalized Floor Mosaic from Pompeii, *AJA* 88, 557-566.

Curtis, R.I. 1984-1986. Product Identification and Advertising on Roman Commercial Amphorae, *Ancient Society* 15-17, 209-228.

Curtis, R.I. 1988a. Spanish Trade in Salted Fish Products in the 1st and 2nd Centuries A.D. *International Journal of Nautical Archaeology and Underwater Exploration* 17, 205-210.
Curtis, R I. 1988b. A. Umbricius Scaurus of Pompeii. *Studia Pompeiana et Classica in Honor of Wilhelmina F. Jashemski*, 1-2. New Rochelle, N.Y., vol. 1, 19-49.
Curtis, R.I. 1991. *Garum and Salsamenta. Production and Commerce in Materia Medica*. Leiden.
Curtis, R.I. 2001. *Ancient Food Technology*. Leiden.
Cuvier, G.L. & M. Valenciennes 1832. Account of the Common Mackerel (Scomber scombrus, Lin.) and the Garum of the Ancients, *Edinburgh Journal of Science* N.S. 12, 286-294.
Danov, C. 1962. Pontos Euxeinos. *RE* Suppl. 9, 866-1175.
Dardaine, S. & J.-N. Bonneville. 1980. La campagne de fouilles d'Octobre 1979 à Belo, *Mélanges de la Casa de Velazquez* 16, 375-419.
D'Arms, J.H. 1981. *Commerce and Social Standing in Ancient Rome*. Cambridge, Mass.
D'Arms, J.H. & E.C. Kopff (eds.) 1980. *The Seaborne Commerce of Ancient Rome: Studies in Archaeology and History*. Rome.
Davidson, J. 1997. *Courtesans and Fishcakes*. London.
de Alarcão, J. (ed.) 1971. *Actas do II Congresso Nacional de Arqueologia (Coimbra, 1970)*, II. Coimbra.
de Alarcão, J. 1988a. *Roman Portugal*, I. Warminster.
de Alarcão, J. 1988b. *Roman Portugal*, II. Warminster.
de Almeida, D.F., J. Cavaleiro Paixão & A. Cavaleiro Paixão. 1978. *Notas sobre a estação arqueológica de Tróia de Setúbal.* Setúbal.
de Figueiredo, A.M. 1906. Ruines d'antiques établissements à salaisons sur le littoral sud du Portugal, *Bulletin Hispanique* 8, 109-121.
de Frutos, G., G. Chic & N. Berriatua. 1988. Las anforas de la factoria prerromana de salazones de "Las Redes" (Puerto de Santa Maria, Cadiz), in: Pereira Menaut (ed.) 1988, 295-306.
Delussu, F. & B. Wilkens 2000. Le conserve di pesce. Alcuni dati da contesti Italiani, *MEFRA* 112, 53-65.
Desse-Berset, N. & J. Desse 2000. *Salsamenta, garum* et autres préparations de poissons. Ce qu'en disent les os, *MEFRA* 112, 84-92.
Dias Diogo, A.M. & A. Cavaleiro Paixão 2001. Ânforas de escavações no provoado industrial romano de Tróia, Steúbal, *Revista portuguesa de arqueologia* 4, 117-40.
Domanskij, J.V. & K.K. Marčenko 2003. Towards Determining the Chief Function of the Settlement of Borysthenes, in: P.G. Bilde, J.M. Højte & V.F. Stolba, *The Cauldron of Ariantas* (Black Sea Studies, 1). Aarhus, 29-36.
Domergue, C., G. Nicolini, D. Nony, A. Bourgeoix, F. Mayet & J.C. Richard 1974. *Excavaciones de la Casa de Velazquez en Belo (Bolonia – Cádiz); Campañas 1966 a 1971* (Excavaciones Arqueologicas en España, 79). Madrid.

Donati, A. & P. Pasini (eds.) 1997. *Pesca e pescatori nell'antichità*. Venice.

Dovatur, A.I., D.P. Kallistov & I.A. Šišova 1982. *Narody našej strany v "Istorii" Gerodota*. Moscow.

Dressel, H. 1879. Di un grande deposito di anfore rinvenuto nel nuovo quartiere del Castro Pretorio, *Bullettino della commissione archeologica communale di Roma* 7, 36-112, 143-195.

Drexhage, H.-J. 1993. Garum und Garumhandel im römischen und spätantiken Ägypten. *MünstBeitr* 12, 27-55.

Dubois, L. 1996. *Inscriptions grecques dialectales d'Olbia du Pont* (Hautes études du monde Gréco-Romain, 22). Genève.

Dumont, J. 1976-77. La pêche du thon à Byzance à l'époque hellénistique, *REA* 78-79, 96-119.

Dupont, P. 2001. Trafics méditerranéens archaiques: quelques aspects, in: R. Eichmann & H. Parzinge (eds.), *Migration und Kulturtransfer. Die Wandel vorderer- und zentralasiatischer Kulturen im Umbruch vom 2. zum 1. vorchristlichen Jahrtausend. Akten des Internationalen Kolloquiums Berlin, 23. bis 26. November 1999*. Bonn, 445-460.

Duval, P.-M. 1949. La forme des navires romains d'après la mosaïque d'Althiburus, *MEFRA* 61, 119-149.

Eberl, G. 1892. *Die Fischkonserven der Alten*. Stadtamhof.

Edmondson, J.C. 1987. *Two Industries in Roman Lusitania: Mining and Garum Production*. (BAR International Series, 362). Oxford.

Edmondson, J.C. 1990. Le *garum* en Lusitanie urbaine et rurale: hiérarchies de demande et de production, in: Gorges (ed.) 1990, 123-147.

Ehmig, U. 1995. Allex oder Anderes, *Mainzer Archäologische Zeitschrift* 2, 117-130.

Ehmig, U. 1996. Garum für den Statthalter. Eine Saucenamphore mit Besitzeraufschrift aus Mainz, *Mainzer Archäologische Zeitschrift* 3, 25-56.

Eiring, J. & J. Lund (eds.) forthcoming. *Transport Amphorae and Trade in the Eastern Mediterranean. Acts of the International Colloquium at the Danish Institute at Athens*, 26-29 September, 2002 (Monographs of the Danish Institute at Athens, 5). Athens.

Empereur, J.-Y. & Y. Garlan 1987. Bulletin archéologique: amphores et timbres amphoriques (1980-1986), *REG* 100, 58-109.

Empereur, J.-Y. & Y. Garlan 1992. Bulletin archéologique: amphores et timbres amphoriques (1987-1991), *REG* 105, 176-220.

Empereur, J.-Y. & Y. Garlan 1997. Bulletin archéologique: amphores et timbres amphoriques (1992-1996), *REG* 110, 161-209.

Esteve Guerrero, M. 1952. Sanlúcar de Barrameda (Cádiz): fábrica de salazón romana en la Algaida, *Noticiario arqueológico hispánico* 1-3, 126-133.

Étienne, R. 1970. À propos du "garum sociorum", *Latomus* 29, 297-313.

Étienne, R. & F. Mayet 1998a. Les mercatores de saumure Hispanique, *MEFRA* 110, 147-165.

Étienne, R. & F. Mayet 1998b. Le *garum* à Pompei. Production et commerce, *REA* 100, 199-215

Étienne, R. & F. Mayet. 1991. Le *garum* à la mode de Scaurus, in: *Alimenta. Estudios en homenaje al Dr. Michel Ponsich.* (Gerion, Anejos 3). Madrid, 187-194.

Étienne, R. & F. Mayet. 2002. *Salaisons et sauces de poisson Hispaniques*. Paris.

Étienne, R., Y. Makaroun & F. Mayet, 1994. *Un grand complexe industriel à Tróia (Portugal)*. Paris.

Faber, G.L. 1883. *Fisheries of the Adriatic and the Fish Thereof: a report of the Austro-Hungarian Sea-Fisheries, with a Detailed Description of the Marine Fauna of the Adriatic Gulf*. London.

Fajen, F. (ed.) 1999. *Oppianus: Halieutica*. Stuttgart.

Fedorov-Davydov, G.A. 1975. O scenach terzanij i bor'by zverej v pamjatnikach skifo-sibirskogo iskusstva, *Uspechi sredneaziatskoj archeologii* 3, 23-28.

Finley, M.I. 1999. *The Ancient Economy*. Updated edition by I. Morris. Berkeley.

Fıratlı, N. & L. Robert 1964. *Les steles funéraires de Byzance gréco-romaine*. Paris.

Flower, B. & E. Rosenbaum 1958. *The Roman Cookery Book*. London.

Foucher, L. 1970. Note sur l'industrie et le commerce des salsamenta et du garum, in: *Actes du 93e Congrès National des Sociétés Savantes. Tours 1968. Section d'Archéologie*. Paris, 17-21.

Francis, R.C., J. Field, D. Holmgren & A. Strom 2001. Historical approaches to the northern California current ecosystem, in: Holm, Smith & Starkey (eds.), 123-139.

Frank, T. 1936. On the Export Tax of Spanish Harbors. *American Journal of Philology* 57, 87-90.

Franke, P.R. 1968. *Kleinasien zur Römerzeit. Griechisches Leben im Spiegel der Münzen*. Munich.

French, R. 1986. Pliny and Renaissance Medicine, in: French, R. & F. Greenaway (eds.) *Science in the Early Roman Empire: Pliny the Elder, His Sources and Influence*. Totowa, N.J.

Gajdukevič, V.F. 1952a. Raskopki Mirmekija v 1935-1938 gg., *MIA* 25, 135-220.

Gajdukevič, V.F. 1952b. Raskopki Tiritaki v 1935-1940 gg., *MIA* 25, 15-134.

Gajdukevič, V.F. 1971. *Das Bosporanische Reich*. Berlin.

Gallant, T.W. 1985. *A Fisherman's Tale* (Miscellanea Graeca, 7). Gent.

Gallant, T.W. 1991. *Risk and Survival in Ancient Greece. Reconstructing the Rural Domestic Economy*. Stanford, Cal.

Garcia y Bellido, A. 1942a. *Fenicios y Cartagineses en Occidente*. Madrid.

García y Bellido, A. 1942b. La industria pesquera y conserva española en la Antigüedad, *Investigación y Progreso* 13, 1-8.

Garlan, Y. (ed.) 1999. *Production et commerce des amphores anciennes en Mer Noire. Colloque international organiseé à Istanbul, 25-28 mai 1994*. Aix-en-Provence.

Garlan, Y. 1990. Remarques sur les timbres amphoriques de Sinope, *Comptes Rendus des Académie des Inscriptions* 1990, 490-507.

Garlan, Y. 2000. *Amphores et timbres amphoriques grecs entre erudition et idéologie*. Paris.

Garlan, Y. 2002. Bulletin archéologique. Amphores et timbres amphoriques (1997-2001), *REG* 115, 149-215.

Garnsey, P. 1998. *Cities, Peasants and Food in Classical Antiquity. Essays in Social and Economic History*. Cambridge.

Garnsey, P. 1999. *Food and Society in Classical Antiquity*. Cambridge.

Gavriljuk, N.A. 1999. *Istorija ekonomiki stepnoj Skifii v VI-III vv. do n.e.* Kiev.

Gavriljuk, N.A., V.P. Bylkova & S.N. Kravčenko 1992. *Skifskie poselenija IV v. do n.e. v stepnom Podneprov'e*. Kiev.

Gavriljuk N.A., V.N. Griščenko & E.N. Jablonovskaja-Griščenko 2001. Ornitofauna v skifskoj torevtike, in: V.Ju. Zuev et al. (eds.), *Bosporskij fenomen*. St. Petersburg, 260-266.

Gavriljuk, N.A. & N.P. Olenkovskij 1992. *Pam'jatki skifiv. Arheologična karta Nyžnodniprovs'kogo regionu*. Cherson.

Gazda, E.K. & A.M. McCann 1987. Reconstruction and function: port, fishery, and villa, in: McCann et al. 1987, 137-159.

Gil Mantas, V. 1999. As *villae* marítimas e o problema do povoamenta do litoral português na época romana, in: Gorges & Rodríguez Martín (eds.) 1999, 135-156.

Gorges, J.-G. 1979. *Les villas hispano-romaines. Inventaire et problématiques archéologiques.* Paris.

Gorges, J.-G. (ed.) 1990. *Les villes de Lusitanie romaine; hiérarchies et territoires. Table ronde internationale du CNRS, Talence, le 8-9 décembre 1988*. Paris.

Gorges, J.-G. & G. Rodríguez Martín (eds.) 1999. *Économie et territoire en Lusitanie romaine*. Madrid.

Gozables Craviota, E. 1997. *Economía de la Mauritania Tingitana (Siglos I A. de C. – II D. de C.).* Ceuta.

Grace, V.R. 1962. Stamped Handles of Commercial Amphoras, in: H.D. Colt (ed.), *Excavations at Nessana (Auja Hafir, Palestine),* 1. London, 106-130.

Grakov, B.N. 1954. *Kamenskoe gorodišče na Dnepre* (*MIA*, 36). Moscow.

Grau Almero, E., G. Pérez Jorda, P. Iborra Eres, J. Rodrigo García, C.G. Rodríguez Santan & S. Carrasco Porras 2001. Gestión de Recursos y Economá, in: Aranegui Gascó (ed.) 2001, 191-230.

Grimal, P. & T. Monod 1952. Sur la véritable nature du "garum", *REA* 54, 27-38.

Gualandri, I. 1967. *Incerti auctori in Oppiani Halieutica paraphrasis.* Milan.

Gudger, E.W. 1924. Pliny's *Historia Naturalis*: the Most Popular Natural History Ever Published, *Isis* 6, 269-281.

Haley, E.W. 1990. The Fish Sauce Trader L. Iunius Puteolanus. *ZPE* 80, 72-78.

Hannestad, L., V.F. Stolba & A.N. Ščeglov (eds.) 2002. *Panskoye I. Vol. 1: The Monumental Building U6*. Aarhus.

Hart, J.B. & J. Reynolds 2002. *Fish Biology* (Handbook of fish biology and fisheries, 1). Oxford.

Head, B.V. 1911. *Historia numorum. A manual of Greek numismatics*. Oxford.

Hesnard, A. 1980. Un dépôt augustéen d'amphores à La Longarina, Ostie, in: D'Arms & Kopff (eds.) 1980, 141-156

Hoddinott, R.F. 1973. *Bulgaria in Antiquity*. New York.

Hörig, M. 1984. Dea Syria. Atargatis, *ANRW* 2.17.3, 1536-1581.

Hoffman, G. & H.D. Schulz 1988. Coastline Shifts and Holocene Stratigraphy on the Mediterranean coast of Andalucia (Southeastern Spain), in: Raban (ed.) 1988, 53-70.

Holm, P., T. Smith, & D. Starkey (eds.) 2001. *The Exploited Seas: New Directions for Marine Environmental History* (Research in Maritime History, 21). St. John's, Newfoundland.

Il'inskaja, V.A. & A.I. Terenožkin 1983. *Skifija VII-IV vv. do n.e.* Kiev.

Immerzeel, M. 1990. *Negotiator Allecarius*. Fabrication et commerce de sauce de poisson dans le Nord-ouest de l'Empire Romain. *Oudheidkundige Mededelingen uit het Rijksmuseum van Oudheiden te Leiden*, 70, 183-192.

Ivanova, N.V. 1994. Fish Remains from Archaeological Sites of the Northern Part of the Black Sea Region (Olbia, Berezan), *Offa* 51, 278-283.

Jacobsen, A.L. Lund 2003. Limfjordens fiskeri 1890-1925, biologi, økonomi og forvaltning. Unpublished dissertation, University of Southern Denmark, Esbjerg.

Jardin, C. 1961. Garum et sauces de poisson de l'antiquité, *Rivista di Studi Liguri* 27, 70-96.

Jefremow, N. 2003. Der Wein vom Taurischen Chersonesos in Thrakien. Zur Geschichte der Handelsbeziehungen zwischen dem Nord- und Westgestade des Pontos Euxeinos während der hellenistischen Zeit, *MünstBeitr* 22, 27-47.

Jennings, S., M. Kaiser & J. Reynolds 2001. *Marine Fisheries Ecology*. Oxford.

Jodin, A. 1957. Note préliminaire sur l'établissement pré-romain de Mogador (campagnes 1956-1957), *Bulletin d'Archéologie Marocaine* 2, 9-40.

Jodin, A. 1967. *Les etablissements du Roi Juba II aux Îles Purpuraires (Mogador)*. Tanger.

Jöhrens, G. 1999. *Amphorenstempel im Nationalmuseum von Athen: zu den von H.G. Lolling aufgenommenen „unedierten Henkelinschriften". Mit einem Anhang: Die Amphorenstempel in der Sammlung der Abteilung Athen des Deutschen Archäologischen Instituts*. Mainz.

Jones, A.K.G. 1988. Fish Bones from Excavations in the Cemetery of St. Mary Bishophill Junior. *The Archaeology of York* 15, 126-131.

Jones, H.L. 1924. *The Geography of Strabo. With an English translation by H.L. Jones.* Cambridge, Mass.

Kac, V.I., S.Y. Monachov, V.F. Stolba & A.N. Ščeglov 2002. Tiles and Ceramic Containers, in: Hannestad, Stolba & Ščeglov (eds.) 2002, 102-126.

Kadeev, V.I. 1962. Rybolovnyj promysel u Chersonesi v peršich vikach n.e., *Učeni zapysky Charkivs'kogo deržavnogo universitetu* 124; *Trudy istoryčnogo fakul'tetu* 9, 59-79.

Kadeev, V.I. 1970. *Očerki istorii ekonomiki Chersonesa Tavričeskogo v I – IV vekach n.e.* Charkov.

Kadeev, V.I. & S.G. Ryzov 1973. Novaja rybozasoločnaja cisterna v Chersonese, *Archeologija Kiev* 12, 76-80.

Karyškovskij, P.O. 1982. Ob izobraženii orla i del'fina na monetach Sinopy, Istrii i Ol'vii, in: V.L. Janin et al. (eds.) *Numizmatika antičnogo Pričernomor'ja.* Kiev, 80-98.

Kastanajan, E.G. 1958. Raskopki Porfmija v 1953 g., *SovA* 3, 203-207.

Kawamura, Y. & M. R. Kare (eds.) 1987. *Umami. A Basic Taste.* New York.

Kbiri Alaoui, M. 2000. *L'établissement punico-maurétanien de Kouass.* Rabat.

Keay, S. 1984. Decline or Continuity? The Coastal economy of the Conventus Terraconensis from the fourth century until the late sixth century, in: Blagg et al. (eds.) 1984, 552-577.

Kent, J.H. 1953. Stamped Amphora Handles from the Delian Temple Estates, in *Studies Presented to David Moore Robinson,* 2. St. Louis, 127-134.

Keydell, Rudolf 1937. Oppians Gedicht von der Fischerei und Aelians Tiergeschichte, *Hermes* 72, 411-434.

King, A. & M. Henig (eds.) 1981. *The Roman West in the Third Century.* Oxford.

Knapp, R.C. & F.H. Stanley, Jr. 2000. Lusitania-Baetica, in Talbert (ed.) 2000, 415-439.

Koehler, C.G. 1996. Wine Amphoras in Ancient Greek Trade, in: P.E. McGovern, S.J. Fleming & S.H. Katz (eds.), *The Origins and Ancient History of Wine.* Amsterdam, 323-337.

Köhler, H.K.E. 1832. Τάριχος, ou recherches sur l'histoire et les antiquités des pêcheries de la Russie méridionale, *Mémoires de l'Academie Impériale de Sciences de St Peterburg*, 6e série, t. 1. St. Petersburg.

Koehne, B. 1857. *Opisanie muzeuma pokojnogo knjazja V.V. Kočubeja i izsledovanija ob istorii i numizmatike grečeskich poselenij v Rossii, ravno kak carstv: Pontijskogo i Bosfora Kimmerijskogo*, 1. St. Petersburg.

Koltuchov, S.G. 1997. Obrazy ryb v greko-varvarskom iskusstve Severnogo Pričernomor'ja, *Archeologija Kryma* 1.1, 59-67.

Koltuchov, S.G., V.A. Kolotuchin & A.E. Kislyj 1994. O rabotach Severo-Krymskoj ekspedicii, in: V.A. Kutajsov (ed.), *Archeologičeskie issledovanija v Krymu. 1993 god.* Simferopol, 153-162.

Korol'kova, E.F. 1998. Ikonografija chiščnoj pticy v skifskom zverinom stile VI-IV vv. do n.e., in: I. Ja. Frojanov et al. (eds.), *Istorija i kul'tura drevnich i srednevekovych obščestv. Problemy archeologii.* St. Petersburg, 166-178.

Kraay, C.M. 1976. *Archaic and Classical Greek Coins.* Berkeley.

Kruglikova, I.T. 1963. Itogi semiletnich raskopok poselenija u d. Semenovki, *KSIA* 95, 43-51.

Kuprin, A.I. 1986. Lestrigony, in: A.I. Kuprin, *Reka žizni: povesti i rasskazy.* Leningrad.

Kutajsov, V.A. 1986. K numizmatike Kerkinitidy 5 v. do n.e., *VDI* 2, 94-97.

Kutajsov, V.A. 1991. Monety Kerkinitidy 5 v. do n.e., *VDI* 1, 46-69.

Kutajsov, V.A. 1995. Cast money and coins of Kerkinitis of the fifth century BC, *Ancient civilizations* 2.1, 39-59.

Kuz'mina, E.E. 1976. O semantike izobraženij na Čertomlyckoj vase, *SovA* 3, 68-75.

Kuz'mina, E.E. 1987. Sjužet bor'by chiščnika i kopytnogo v iskusstve "zverinogo" stilja Evrazijskich stepej skifskoj epochi, in: A.I. Martynov & V.I. Molodin (eds.) *Skifo-sibirskij mir: iskusstvo i ideologija.* Novosibirsk, 3-12.

Lagóstena Barrios, L. 2001. *La producción de salsas y conservas de pescado en la Hispania Romana (II a.C – VI d.C).* Barcelona.

Lawall, M.L. 1995. Transport amphoras and trademarks: imports to Athens and economic diversity in the fifth century B.C. Ph.D. dissertation, University of Michigan.

Lawall, M.L. 1998. Ceramics and positivism revisited: Greek transport amphoras and history, in: H. Parkins & C. Smith (eds.), *Trade, Traders and the Ancient City.* London, 75-101.

Lawall, M.L. 1999. Studies in Hellenistic Ilion: Transport Amphoras from the Lower City, *Studia Troica* 9, 187-224.

Lawall, M.L. 2000. Graffiti, Wine Selling, and the Reuse of Amphoras in the Athenian Agora, ca. 430 to 400 B.C., *Hesperia* 69, 3-90.

Lebedev, V.D. & Ju. E. Lapin 1954. K voprosu o rybolovstve v Bosporskom carstve, *MIA* 33, 197-214.

Lepiksaar, J. 1973. Fischknochenfunde aus der phönizischen Faktorei von Toscanos, in: Boessneck (ed.) 1973, 109-119.

Lernau, O., H. Cotton & Y. Goren 1996. Salted Fish and Fish Sauces from Masada: A Preliminary Report, *Archaeofauna* 5, 35-41.

Leskov, O.M. 1972. *Skarby kurganiv Chersonščiny.* Kiev.

Lestocquoy, J. 1952. Épices, médicine et abbayes, in: *Études mérovingiennes. Actes des journées de Poitiers, 1-3 Mai 1952.* Paris, 179-186.

Lindberg, G.U. 1971. *Families of the fish of the world. A check list and a key.* Leningrad.

Liou, B. 1982. Informations archéologiques: Corse du Sud, *Gallia* 40, 439-444.

Liou, B. & R. Marichal 1978. Les inscriptions peintes sur amphores de l'anse Saint-Gervais à Fos-sur-mer, *Archeonautica* 2, 165, No. 169.

Liou, B. & E. Rodríguez Almeida 2000. Les inscriptions peintes des amphores du Pecio Gandolfo (Almería). *MEFRA* 112, 7-25.

Litvinskij, B.A. 1975. Pamirskaja kosmologija, *Strany i narody Vostoka* 16, 253-257.

Lomtadze, G. & D. Zhuravlev forthcoming. Amphorae from a Late Hellenistic Cistern at Pantikapaion, in Eiring & Lund (eds.) forthcoming.

Lopetcharat, K., Y. J. Choi, J. W. Park & M. A. Daeschel 2001. Fish Sauce Products and Manufacturing, *Food Reviews International* 17, 65-88.

Lowe, B.J. 1997. The Trade and Production of Garum and its Role in the Provincial Economy of Hispania Tarraconensis. Ph.D. dissertation, Edinburgh.

Loza Azuaga, M.L. & J. Beltrán Fortes. 1988. Estudio arqueológico del yacimiento romano de Haza Honda (Málaga), in: Ripoll Perelló (ed.) 1988, 991-1001.

Lund, J. 2002. Olie på vandene? in: L.K. Jacobsen & A.M. Carstens (eds.), *Til Jens Erik Skydsgaard* (Meddelelser fra Klassisk Arkæologisk Forening, Suppl. 1), 15-21.

Lund, J. forthcoming. Oil on the Waters? Reflections on the Contents of Hellenistic Transport Amphorae from the Aegean in: Eiring & Lund (eds.) forthcoming.

Mabesoone, J.M. 1963. Observations on the sedimentology and geomorphology of the Guadalete drainage area (Cádiz, Spain), *Geologi Mijnbouw* 42, 309-328.

McCann, M., J. Bourgeois, E.K. Gazda, J.P. Oleson & E.L. Will 1987. *The Roman port and fishery of Cosa; a center of ancient trade*. Princeton, N.J.

Mack, G.R. (ed.) 2003. *Crimean Chersonesos. City, Chora, Museum, and Environs*. Austin, Texas.

MacKendrick, P., 1980. *The North African Stones Speak*. Chapel Hill, N.C.

Mackie, I. M., R. Hardy & G. Hobbs. 1971. *Fermented Fish Products* (FAO Fisheries Reports, 100). Rome.

Mancevič, A. 1987, in: A.Ju. Alekseev, *Kurgan Solocha*. Leningrad.

Marčenko, K.K, V.G. Žitnikov & V.P. Kopylov 2000. *Die Siedlung Elizavetovka am Don*. Moscow.

Marlière, É. 2002. *L'outre et le tonneau dans l'Occident romain*. Montignac.

Marquardt, J. 1892. *La vie privée des Romains* (Manuel des antiquités Romaines, 15). Paris.

Marques da Costa, A.I. 1930-31. Estudos sobre algumas estações da época luso-romana nos arredores de Setúbal, *O Arqueologo Português* 29, 2-31.

Marques da Costa, J. 1960. *Novos elementos para a localização de Cetobriga*. Setúbal.

Marti, Ju. 1941. Pozdne-ellinističeskie nadgrobija Bospora kak istoriko-kul'turnyj dokument, *SovA* 7, 31-44.

Marti, Ju. 1941a. Novye dannye o rybnom promysle v Bospore Kimmerijskom po raskopkam Tiritaki i Mirmekija, *SovA* 7, 94-106.

Marti, Ju. 1941b. Rybozasoločnye vanny Tiritaki, *MIA* 4, 93-95.
Martin, G. 1970. Las pesquerías romanas de la costa de Alicante, *Saguntum: Papeles del laboratorio de arquelogía de Valencia* 10, 139-153.
Martin, G. & M.D. Serres. 1970. *La factoría pesquera de Punta de l'Arenal y otros restos romanos de Jávea (Alicante).* Valencia.
Martin-Bueno, M., J. Liz Guiral & M.-L. Cancela Ramirez de Arellano 1984. Baelo Claudia: Sector Sur 1981-1983 (Avance), *Mélanges de la Casa de Velazquez* 20, 487-496.
Martin-Kilcher, S. 1987. *Die römischen Amphoren aus Augst und Kaiseraugst. Ein Beitrag zur römischen Handels- und Kulturgeschichte. 1: Die südspanischen Ölamphoren (Gruppe 1)* (Forschungen in August, 7/1). Augst.
Martin-Kilcher, S. 1990. Fischsaucen und Fischkonserven aus dem römischen Gallien. *Archäologie der Schweiz* 13, 37-44.
Martin-Kilcher, S. 1994a. *Die römischen Amphoren aus Augst und Kaiseraugst. Ein Beitrag zur römischen Handels- und Kulturgeschichte. 2: Die Amphoren für Wein, Fischsauce, Sudfrüchte (Gruppen 2-24)* (Forschungen in August, 7/2). Augst.
Martin-Kilcher, S. 1994b. *Die römischen Amphoren aus Augst und Kaiseraugst. Ein Beitrag zur römischen Handels- und Kulturgeschichte. 3: Archäologische und Naturwissenchaftliche Tonbestimmungen* (Forschungen in August, 7/3). Augst.
Masanov, N.E. 1989. Tipologija skotovodčeskogo chozjajstva kočevnikov Evrazii, in: *Vzaimodejstvie kočevyh kul'tur i drevnih civilizacij.* Alma-Ata, 55-81.
Mehl, A. 1987. Der Überseehandel von Pontos, in: *Stuttgarter Kolloquium zur historischen Geographie des Altertums 1 1980,* Bonn, 103-186.
Michel, S. 1995. *Der Fisch in der skythischen Kunst* (Europäische Hochschulschriften, 38.52). Frankfurt.
Minns, E.H. 1913. *Scythians and Greeks.* Cambridge.
Molina, F. & C. Huertas 1985. *Almuñécar en la Antigüedad,* II. Granada.
Mongait, A. 1959. *Archaeology in the U.S.S.R.* Moscow.
Muñoz Vicente, Á., G. de Frutos Reyes & N. Berriatua Hernández 1988. Contribución a las orígenes y difusión comercial de la industria pesquera y conservera Gaditana a través de las recientes aportaciones de las factorías de salazones de la Bahía de Cadiz, in: Ripoll Perelló (ed.) 1988, 487-508.
Mušmov, N.A. 1912. *Antičnite monety na Balkanskija poluostrov i monetite na bulgarskite care.* Sofia.
Naster, P. 1959. *Catalogue des monnaies grecques. La collection Lucien de Hirsch.* Bruxelles.
Nicolaou, K. & A. Flinder 1976. Ancient fish-tanks at Lapithos, Cyprus, *International Journal of Nautical Archaeology and Underwater Exploration* 5.2, 133-141.
Nikol'skij, G.V. 1937. K poznaniju ichtiofauny r. Kubani, *Bjulleten' Moskovskogo obščestva ispytatelej prirody. Otdelenie biologii* 45.2, 121-124.

Nock, A.J. & C.R. Wilson (eds.) 1931. *The Works of Francis Rabelais* I-II. New York.

Nolla-Brufau, J.M. 1984. Excavaciones recientes en la ciudadela de Roses. El edificio Bajo-imperial, in: Blagg et al. (eds.) 1984, 430-459.

Ørsted, P. 1998. Salt, fish and the Sea in the Roman Empire, in: I. Nielsen & H.S. Nielsen (eds.), *Meals in a Social Context. Aspects of the Communal Meal in the Hellenistic and Roman World* (Aarhus Studies in Mediterranean Antiquity, 1). Aarhus, 13-35.

Olson, S. D. & A. Sens 2000. *Archestratos of Gela. Greek Culture and Cuisine in the Fourth Century BCE*. Oxford.

Onajko, N.A. 1956. O fanagorijskich gruzilach, *MIA* 56, 154-163.

Onajko, N.A. 1976a. O vozdejstvii grečeskogo iskusstva na meoto-skifskij zverinyj stil', *SovA* 3, 76-86.

Onajko, N.A. 1976b. Zverinyj stil' i antičnyj mir Severnogo Pričernomor'ja v VII – IV vv. do n.e., in: A.I. Meljukova & M.G. Moškova (eds.) *Skifo-sibirskij zverinyj stil' v iskusstve narodov Evrazii*. Moscow, 66-73.

Orešnikov, A.V. 1892. *Materialy po drevnej numizmatike Černomorskogo poberež'ja*, Moscow.

Ostapenko, I.A. 2001. Pam'jatki osilosti skifskogo času na ostrovi Chortica, *Arheologia Kiev* 1, 51-68.

Pack, R. (ed.) 1963. *Onirocriticon Libri V*. Leipzig.

Panella, C. 1972. Annotazioni in margine alle stratigrafie delle Terme ostiensi del Nuotatore, *Recherches sur les amphores romaines. Collection de l'École Française de Rome*, 10. Rome, 151-165.

Paoli, U. 1975 (1940). *Rome, Its People, Life and Customs*. Trans. by R.D. McNaughton. Florence.

Parker, A.J. 1977. Lusitanian amphoras, in: *Méthodes classiques et méthodes formelles dans l'étude des amphores. Actes du Colloque de Rome, 27-29 Mai 1974* (Collection de l'École Française de Rome, 32), Paris, 35-46.

Parker, A.J. 1992. *Ancient Shipwrecks of the Mediterranean and the Roman Provinces* (BAR International Series, 580). Oxford.

Parkins, H.M. (ed.) 1997. *Roman Urbanism. Beyond the Consumer City*. London.

Pashkevich, G.A. 2001. Archaeobotanical studies on the northern coast of the Black Sea, *Eurasia Antiqua* 7, 511-67.

Paterson, J. 1998. Trade and Traders in the Roman World: Scale, Structure, and Organisation, in: H. Parkins & C. Smith (eds.) *Trade, Traders and the Ancient City*. London, 149-167.

Peacock, D.P.S. 1974. Amphorae and the Baetican Fish Industry, *The Antiquaries Journal* 54, 232-243.

Peacock, D.P.S. 1977. Roman Amphorae: Typology, Fabric and Origins, in *Méthodes classiques et méthodes formelles dans l'étude des amphores* (Collection de l'École Française de Rome, 32). Rome, 261-278.

Peacock, D.P.S. & D.F. Williams 1985. *Amphorae and the Roman Economy: an Introductory Guide*. London and New York.
Peacock, D.P.S. & D.F. Williams 1991. *Amphorae and the Roman Economy: an introductory guide* (paperback edition). London.
Pekáry, I. 1999. *Repertorium der hellenistischen und römischen Schiffsdarstellungen* (*Boreas*, Beiheft 8). Münster.
Pelletier, A. 1988. Belo: une cite romaine du détroit, in: Ripoll Perelló (ed.) 1988, 801-810.
Pellicer Catalán, M. 2002. Phoenician and Punic Sexi, in: Bierling (ed.) 2002, 49-77.
Pereira Menaut, G. (ed.) 1988. *Actas 1ero Congreso Peninsular de Historia Antigua,* 1. Santiago de Compostela.
Pogrebova, N.N. 1958. Pozdneskifskie gorodišča na Nižnem Dnepre (gorodišča Znamenskoe i Gavrilovskoe), in: K.F. Smirnov (ed.) *Pamjatniki skifo-sarmatskoj archeologii v Severnom Pričernomor'e* (*MIA*, 64). Moscow, 103-247.
Poinssot, Cl. 1965. Quelques remarques sur les mosaïques de la maison de Dionysos et d'Ulysse à Thugga (Tunisie), in: *La mosaïque greco-romaine. (Colloque International, Paris 29 Août – 3 Septembre 1963).* Paris, 219-232.
Polos'mak, N. 2001. *Vsadniki Ukoka*. Novosibirsk.
Ponsich, M. 1967. Kouass, port antique et carrefour des voies de la Tingitane, *Bulletin d'Archéologie Marocaine* 7, 369-405.
Ponsich, M. 1968. *Alfarias de época fenicia y púnico-mauritana en Kuass (Arcila, Marruecos).* Valencia.
Ponsich, M. 1970. *Recherches archéologiques a Tanger et dans sa région*. Paris.
Ponsich, M. 1975. Pérennité des relations dans le circuit de Détroit de Gibraltar, *ANRW* 2.3, 655-684.
Ponsich, M. 1976. A propos d'une usine antique de salaisons à Belo (Bolonia-Cadix), *Mélanges de la Casa de Velazquez* 12, 69-79.
Ponsich, M. 1981. *Lixus: Le quartier des temples (etude preliminaire).* Rabat.
Ponsich, M. 1988. *Aceite de oliva y salazones de pescado. Factores geo-economicos de Betica y Tingitania*. Madrid.
Ponsich, M. & M. Tarradell 1965. *Garum et industries antiques de salaison dans la Méditerranée occidentale.* (Bibliothèque de l'École des Hautes Études Hispaniques, 36). Paris.
Powell, O. 2003. *Galen. On the Properties of Foodstuffs*. Cambridge.
Purpura, G. 1982. Pesca e stabilimenti antichi per la lavorazione del pesce in Sicilia: I. S. Vito (Trapani), Cala Minnola (Levanzo), *Sicilia Archeologica* 48, 45-60.
Raban, A. (ed.) 1988. *Archaeology of Coastal Changes; Proceedings of the First International Symposium "Cities on the Sea – Past and Present" Haifa, Israel, September 22-29, 1986.* Oxford.
Raevskij, D.S. 1977. *Očerki ideologii skifo-sakskich plemen*. Moscow.
Raevskij, D.S. 1985. *Model' mira skifskoj kul'tury. Problemy mirovozzrenija iranojazyčnych narodov evrazijskich stepej I tysjačeletija do n.e.* Moscow.

Rau, V. 1984. *Estudos sobre a história do sal português*. Lisbon.
Rebuffat, R. 1972. Les fouilles de Thamusida et leur contribution à l'histoire du Maroc, *Bulletin d'Archéologie Marocaine* 8, 51-65.
Rebuffat, R. 1977. *Thamusida*, III. Rome.
Reece, R. 1981. The Third Century: Crisis or Change? in: King & Henig (eds.) 1981, 27-38.
Remarck, P. 1912. *De Amphorarum Inscriptionibus Latinis Quaestiones Selectae*. Tübingen.
Rhode, P. 1890. *Thynnorum Captura Quanti Fuerit apud Veteres Momenti*. Leipzig.
Ribeiro, M. 1971. Anzois de Troia. Subsudios para o estudo da pesca no periodo lusitano-romano, in: de Alarcão (ed.) 1971, 391-402.
Ripoll López, S. 1988. El atún en las monedas antiguas del estrecho y su simbolismo económico y religioso, in: Ripoll Perelló (ed.) 1988, 481-486.
Ripoll Perelló, E. (ed.) 1988. *Actas de Congreso Internacional el Estrecho de Gibraltar, Ceuta – Noviembre 1987*, I. Madrid.
Romančuk, A.I. 1973. Novye materialy o vremini stroitel'stva rybozasoločnych cistern v Chersonese, *Antičnaya drevnost' i srednie veka* 9, 45-53.
Romančuk, A.I. 1977. Plan rybozasoločnych cistern v Chersonese, *Antičnaya drevnost' i srednie veka* 14, 18-20.
Rondelet, G. 1554-1555. *Libri De Piscibus Marinis*. Lyon.
Rostovcev, M.I. 1913. Predstavlenie o monarchičeskoj vlasti v Skifii i na Bospore, *IAK* 49, 1-62.
Rostovtzeff, M.I. 1929. *The Animal Style in South Russia and China* (Princeton monographs in Art and Archaeology, 14). Princeton, N.J.
Rostovtzeff, M.I. 1941. *The Social and Economic History of the Hellenistic World*. Oxford.
Rostovtzeff, M.I. 1957 (1926). *The Social and Economic History of the Roman Empire*, 2nd ed. Oxford.
Rudenko, S.I. 1953. *Kul'tura naselenija Gornogo Altaja v skifskoe vremja*. Moscow.
Rudenko, S.I. 1960. *Kul'tura naselenija Centralnogo Altaja v skifskoe vremja*. Moscow and Leningrad.
Ruiz Mata, D. 2002. The Ancient Phoenicians of the 8th and 7th Centuries B.C. in the Bay of Cádiz: State of the Research, in: Bierling (ed.) 2002, 155-198.
Rusjaeva, A.S. 1992. *Religija i kul'ty antičnoj Ol'vii*. Kiev.
Sahrhage, D. & Lundbeck, J. 1991. *A History of Fishing*. Hamburg.
Ščeglov, A.N. 1969. Novyj metod opredelenija veličiny ryb po češue i nekotorye dannye o promysle kefali v Severo-Zapadnom Krymu v 1 v. do n.e., *KSIA* 119, 128-130.
Ščeglov, A.N. 1978. *Severo-Zapadnyj Krym v antičnuju epochu*. Leningrad.
Ščeglov, A.N. 2002. Monumental Building U6, in: Hannestad, Stolba & Ščeglov (eds.), 29-98.

Ščeglov, A.N. & V.D. Burdak 1965. O promysle ryby u beregov Tarchankuta v antičnuju epochu, *Rybnoe chozjajstvo* 3, 21-23.
Schönert-Geiss, E. 1970. *Die Münzprägung von Byzantion*, I. *Autonome Zeit* (Griechisches Münzwerk. Schriften zur Geschichte und Kultur der Antike, 2). Berlin.
Schönert-Geiss, E. 1972. *Die Münzprägung von Byzantion*, II. *Kaiserzeit* (Griechisches Münzwerk. Schriften zur Geschichte und Kultur der Antike, 2). Berlin.
Semenov-Zuser, S. 1947. Rybnyj rynok v Chersonese (Marché des poissons à Chersonèse Taurique), *VDI* 1947:2, 237-246.
Semenov-Zuser, S.A. 1947a. *Rybnoe chozjastvo i rynki na juge SSSR v drevnosti. Tabula marmorea universitatis Charcovensis* (Pamjatnikiistorii ekonomiki ikul'tury drevnego Severnogo Pričernomor'ja, 1). Charkov.
Semenov-Zuser, S.A. 1947b. Rybnyj rynok v Chersonese, *VDI* 1947, 2, 237-246.
Šelov, D.B. 1970. Tanais i Nišnij Don v III-I vv. do n.e. Moscow.
Shelov, D.B. 1978. *Coinage of the Bosporus VI-II centuries BC* (BAR International Series, 46). Oxford.
Smidth, J.K. 1875. Historical Observations on the Condition of the Fisheries among the Ancient Greeks and Romans, and on their Mode of Salting and Pickling Fish. (U.S. Congress, Senate. 43rd Congress Miscellaneous Documents, 108). Washington. [Also published in Danish in *Tidsskrift for Fiskeri*, 1871, 34-62].
Smith, T. 1994. *Scaling Fisheries.* Cambridge.
Soares, J. 1980. *Estação romana de Tróia*. Setúbal.
Sokol'skij, N.I. 1968. Svjatilišč e Afrodity v Kepach, *SovA* 1968, 101-118.
Sotomayor, M. 1971. Nueva factoría de salazones de pescado en Almuñécar (Granada), *Noticiario arqueológico hispánico* 15, 147-178.
Sternberg, M. 2000. Donnée sur les produits fabriqués dans une officine de Neapolis (Nabeul, Tunisie), *MEFRA* 112, 133-153.
Stoddart, J.L. 1850. On the Inscribed Pottery of Rhodes, Cnidus, and Other Greek Cities, *Transactions of the Royal Society of Literature of the United Kingdom*, 2nd Series, 3, 1-127.
Stoddart, J.L. 1853. On the Lettered Vase-Stamps from Greek Cities of the Mediterranean and Euxine seas, *Transactions of the Royal Society of Literature of the United Kingdom*, 2nd Series, 4, 1-67.
Stolba, V.F. 1991. Dom IV v. do n.e. na poselenii Panskoe I (raskopki 1987 g.), *KSIA* 204, 78-84.
Stolba, V.F. 2002. Graffiti and dipinti, in: Hannestad, Stolba & Ščeglov (eds.), 228-244.
Struck, M.L. 1912. *Die antiken Münzen von Thrakien.* I, 1. *Die Münzen der Thraker und der Städte Abdera, Ainos, Anchialos* (Die antiken Münzen Nord-Griechenlands, 2). Berlin.

Surov, E.G. 1948. Chersonesskie cisterny. (Po archivnym dannym), *Učenye zapiski Sverdlovskogo gosudarstvennogo pedagogičeskogo instituta* 4, 3-47.

Sutherland, C.H.V. 1939. *The Romans in Spain*. London.

Tailliez, P. 1961. Travaux de l'été 1958 sur l'épave de "Titan" à l'île de Lévant (Toulon), in: *Atti del II Congresso Internazionale di Archaeologia Sottomarina, Albenga 1958*. Bordighera, 173-198.

Talbert, R. (ed.) 2000. *The Barrington Atlas of the Greek and Roman World*. Princeton.

Tarradell, M. 1955. La crisis del siglo III de J.-C. en Maurruecos, *Tamusida* III, 75-105.

Tarradell, M. 1968. *Economía de la colonización fenica*. Barcelona.

Tavares da Silva, C. 1980. *Escavações arqueologicas na Praça do Bocage: 2000 anos de historia*. Setúbal.

Tavares da Silva, C. & J. Soares. 1993. *Ilha do Pessegueiro. Porto romano da costa alentejana*. Lisbon.

Terenožkin, A.I., V.A. Il'inskaja, E.V. Černenko & B.N. Mozolevskij 1973 Skifskie kurgany Nikopol'ščiny, in: A.I. Terenožkin (ed.), *Skifskie drevnosti*. Kiev, 113-187.

Testaguzzi, O. 1970. *Portus*. Rome.

Thayer, P. 2001ff. Atlas of Roman Pottery. http://www.potsherd.uklinux.net/. Cited March 2003.

Tichij, M. 1917. *Ančous Chersonesa Tavričeskago*. (Abstract from the *Vestnik rybopromyšlennosti* 1-3, 1917). Petrograd.

Tokarev S.A. (ed.) 1998. *Mify narodov mira* (Encyklopedija, 1). Moscow.

Tolstoj, I.I. 1953. *Graffiti grečeskich gorodov Severnogo Pričernomor'ja*. Moscow.

Toporov, V.N. 1972. K proischoždeniju nekotorych poetičeskich simvolov. Paleolitičeskaja epocha, in: E.M. Meletinskij (ed.), *Rannie formy iskusstva*. Moscow, 77-103.

Tozer, H.F. 1893. *Selections from Strabo. With an introduction on Strabo's life and works*. Oxford.

Tsetskhladze, G.R. 1998. Trade on the Black Sea in the archaic and classical periods: some observations, in: H. Parkins, H. & C. Smith (eds.), *Trade, Traders and the ancient City*. London, 52-74.

Uerpmann, M. 1972. Archäologische Auswertung der Meeresmolluskenreste aus der west-phönizischen Faktorei Toscanos, *Madrider Mitteilungen* 13, 164-171.

Van Neer, W. & A. Lentacker 1994. New Archaeological Evidence for the Consumption of Locally-produced Fish Sauce in the Northern Provinces of the Roman Empire, *Archaeonautica* 3, 53-62.

Van Veen, A.G. 1965. Fermented and Dried Seafood Products in Southeast Asia, in: G. Borgström (ed.), *Fish as Food*, 3. Paris, 227-250.

Vasmer, M. 1923. *Untersuchungen über die ältesten Wohnsitze der Slaven*, I: *Die Iranier in Südrussland*. Leipzig.

Vinokurov, N.I. 1994. Rybozasoločnye kompleksy chory Evropejskogo Bospora, *RosA* 1994, 4, 154-170.

Voskresensky, N.A., 1965. Salting of Herring, in: G. Borgström (ed.), *Fish as Food*, 3. Paris, 107-131.

Waddington, W.H., E. Babelon & Th. Reinach 1904. *Recueil général des monnaies grecques d'Asie Mineure.* I, 1. *Pont et Paphlagonie*. Paris.

Waelkens, M., W. Van Neer, B. De Cupere & I. Beuls 2003. Hunting and stockbreeding, in: Vanhaverhelke, H. & M. Waelkens, *The Chora of Sagalassos. The Evolution of the Settlement Pattern from Prehistoric until Recent Times* (Studies in Eastern Mediterranean Archaeology, 5). Leuven, 58-60.

Whittaker, C.R. 1994. *Frontiers of the Roman Empire: A Social and Economic Study*. London.

Wilkins, J. 2000. *The Boastful Chef: The Discourse of Food in Ancient Greek Comedy*. Oxford.

Wilkins, J. 2001. Les poissons faisaient-ils partie de la diète ancienne? in: J.-N. Corvisier & M. Bellancourt (eds.), *Démographie et médicine*. Arras, 183-191.

Will, E.L. 2001. Truth in Roman Labeling? *AJA* 105, 263.

Williams, C.K. 1979. Corinth, 1978: Forum Southwest, *Hesperia* 48, 105-144.

Wright, G.R.H. 1990. Of fish and men. Fish symbols in ancient religion, *JPrehistRel*, 3-4, 30-44.

Yamaguchi, S. & K. Ninomiya. 1998. What is Umami? *Umami*. Special issue of *Foods Reviews International* 14, 123-138.

Zahn, R. 1910. Garum, *RE* 7.1, 841-849.

Zeepvat, R.J. 1988. Fishponds in Roman Britian, in: Aston (ed.) 1988, 17-26.

Zelenin, D.K. 1991. *Vostočnoslavjanskaja etnografija*. Moscow.

Zevi, F. 1966. Appunti sulle anfore romane, *Archeologia classica* 18, 208-247.

Žitnikov, V.D. 1992. *Nižnee Podon'e v 6 – pervoj treti 3 vv. do n.e. (ekonomičeskaja charakteristika). Avtoreferat dissertacii*. Moscow.

Zograf, A.N. 1951. *Antičnye monety* (*MIA*, 16). Moscow.

Zograph, A.N. 1977. *Ancient coinage* (BAR International series, 33). Oxford.

Zolotarev, M.I. 1986. Novye materialy o vzaimootnošenijach Ol'vii i Zapadnogo Kryma v 6 – 5 vv. do n.e., *VDI* 1986.2, 88-93.

Contributors

Tønnes Bekker-Nielsen
University of Southern Denmark, Esbjerg
Niels Bohrs vej 9
DK-6700 Esbjerg
Denmark
e-mail: tonnes@hist.sdu.dk

Robert I. Curtis
Department of Classics
Park Hall
University of Georgia
Athens, Georgia 30602-6203
e-mail: ricurtis@uga.edu

Bo Ejstrud
University of Southern Denmark, Esbjerg
Niels Bohrs vej 9
DK-6700 Esbjerg
Denmark
e-mail: ejstrud@hist.sdu.dk

Vincent Gabrielsen
Department of History
University of Copenhagen
Njalsgade 80
DK-2300 København S
Denmark
e-mail: gabriels@hum.ku.dk

Nadežda Gavriljuk
Institute of Archaeology
Ukrainian National Academy of Sciences
Geroev Stalingrada av. 12
UKR-04211 Kiev
Ukraine
e-mail: gavr@iananu.kiev.ua

Jakob Munk Højte
Danish National Research Foundation's
Centre for Black Sea Studies
University of Aarhus
Nordre Ringgade 1
DK-8000 Aarhus C
Denmark
e-mail: klajh@hum.au.dk

Anne Lif Lund Jacobsen
University of Southern Denmark,
Esbjerg
Niels Bohrs vej 9
DK-6700 Esbjerg
Denmark
e-mail: alj@fimus.dk

John Lund
Collection of Classical and Near Eastern
Antiquities
The National Museum of Denmark
Frederiksholms Kanal 12
DK-1220 København K
Denmark
e-mail: john.lund@natmus.dk

Vladimir F. Stolba
Danish National Research Foundation's
Centre for Black Sea Studies
University of Aarhus
Nordre Ringgade 1
DK-8000 Aarhus C
Denmark
e-mail: klavs@hum.au.dk

Athena Trakadas
Department of Classical Archaeology
University of Aarhus
Nordre Ringgade 1
DK-8000 Aarhus C
Denmark
e-mail: athena@maritimehistory.org

John Wilkins
University of Exeter
Queen's Building
The Queen's Drive
Exeter UK EX4 4QH
United Kingdom
e-mail: j.m.wilkins@exeter.ac.uk

Index of Persons

Index of Places

Index of Ancient Sources